James Carpenter

Forging Change

Agile Restructuring in Practice

FIRST EDITION (VERSION 1.1)

Published by Agile Carpentry

Forging Change— Agile Restructuring in Practice
James Carpenter

Version 1.1

Copyeditor: DeAnna Burghart
Cover Design: Deb Tremper
Interior Design: James Carpenter

Published in the United States by Agile Carpentry (http://agilecarpentry.com)

ISBN 978-1-7328751-0-4

Contents

For my wife
Deirdre Carpenter

My parents
Joe & Marjory Carpenter

And my grandmother
Sue Carpenter

Foreword

To paraphrase Ronald Reagan, it isn't so much that managers are ignorant. It's just that they know so many things that aren't so. We build our organizational structures on assumptions that, because of changes in the type of work we do or because they were never true to begin with, are provably false.

Agile is one response to this disconnect. Some of those who experienced the problems so common in large organizations developed alternative approaches—understanding, at least intuitively, that there must be a better way.

One of the most fascinating aspects of agile and the many related frameworks is the deep set of knowledge they are based upon. This knowledge is extensive and well researched. Unfortunately, many if not most managers in a position to design an organizational structure and see it implemented are not well versed in this subject. More difficult still is the fact that many large companies seem, on first glance, to be doing quite well. If money is flowing in, why change? It is the age-old problem: degrees of separation changing data into story. The employees understand exactly what's happening but they exist far from the tip of the pyramid where the power lies. Each level in the pyramid makes adjustments to the data, all for pragmatic reasons, and in the end the truth gets hidden under an avalanche of equivocation.

At the heart of agile is the attempt to build in the adaptability needed when things do change and the old model stops working. This change can be environmental or internally generated as a result of years—decades—of deterioration in the organization. Far too often organizations wait until the signs of change are obvious, at which point it's probably too late to change to meet the new challenge. This is essentially the unstated bet many managers are making—that they can change sufficiently, and quickly enough, after the new facts on the ground are incontrovertible. The euphemism "fast follower" refers to a firm that doesn't need to be on the bleeding edge, but will allow other companies to chart a course they can follow. However, even following requires adaptability. No company has unlimited time or resources to adapt, and yet that is the implicit assumption upon which many managers build their strategy. This isn't a good bet to make.

The question is, what do companies do about it before the change is required—how do managers shift focus to the methods that are known to produce better outcomes and more adaptable organizations? Implementing agile practices and organizational structures is an obvious answer—and the one currently in vogue. This new buzz has positive and negative implications for the movement. On one hand, more people are hearing about other ways of working, and that can only be seen as a positive. On the other hand, there is the same issue any new business fad faces: in order to go broad in appeal, the essence of the change is perverted to avoid the difficult bits.

Fortunately, perhaps, the actual mechanics of "being agile" are quite simple. I say "perhaps" because it's not always clear that simplicity makes things simpler to do. Often, simple, well-founded advice is the most difficult to put into practice because it isn't overloaded with the theatrics so often associated with large-scale change efforts. At its core, agile is about how people in complex environments work best, and that model is fairly simple to explain.

The delta between knowledge and practice—or rather, what people think they know versus what really works in practice—can be vast. Many of the consultants and experts in agile are simply selling something. They will often sell whatever meets with the desire of the manager paying the bill. All too often this is merely a veneer of change and not real, fundamental change in the organization. This isn't a problem unique to agile adoptions; it afflicts any change initiative.

James and I met, as many professionals do today, online. I was on the hunt for an agile coach to help with a large-scale transformation and had asked my contacts for recommendations. James was unique in that he provided examples and documents on his thinking. This is rarer than one might assume. Save for a few in the field who've published books, many consultants just talk, with little documentary evidence of what they believe. This is no doubt because they are selling something and it's best not to prejudice oneself before the sale is made. Much of what James provided was the material that would eventually become this book. I appreciated the opportunity to better understand what he believes without the normal dance of a buyer and seller, where inevitably the seller feels compelled to tell the buyer what they want to hear. James's views of organizational change are rooted in both his personal experience as a developer and a study of the available research. This also is, unfortunately, rarer than one might assume.

In *Forging Change*, James provides an overview of the changes necessary for an agile adoption, with a particular focus on the teams and on specific advice informed by his years of experience. Although an agile adoption is an organizational change process, the work really only happens at the team level—without functioning teams an organization is unlikely to benefit from any kind of agile change effort. Some of the material in this book may be difficult to digest for some managers. As with any change effort, the discomfort is an indication that real change is happening.

If you are looking to change and adapt your organization for the future, be thankful you have the opportunity to learn from those who have been through change and know what's possible. It won't eliminate the difficulty of the journey you're about to embark on, but it will make it a bit more scenic.

David Stackleather

Introduction

If you are actively trying to evolve your organization into one which more fully embraces agile culture and practices, you deserve to hear the brutal truths without any sugar coating. In part 1 of this book I attempt to describe these truths as clearly as possible. I then go on to provide actionable guidance and conceptual models which can be used to achieve positive, lasting organizational change. I hope this will help you to more easily detect and articulate problems in your own organization's process and expectations.

Part 2 provides a loosely organized set of techniques, examples, and references you should find useful when practicing agile methods within your organization. In many cases a few paragraphs coupled with a few diagrams and tables are all that is needed to act as an effective reference for a topic.

I do not intend to provide yet another introductory book on agile process and techniques. Rather, I am trying to cut through the noise that often surrounds large-scale agile organizational transformations. Part 1 does this by reframing the problem from an actionable perspective. Part 2 is intended to help quickly establish concepts and terminology consistent with that new perspective.

I have provided a graphical chapter index at the top of most pages. I hope this will make quick reference easier while helping you discover chapters of interest.

A collection of chapter-specific reference content is available at http://forgingchange.com. The relevant link, along with a QR code to the same, is provided at the end of each chapter.

About the Author

James Carpenter started life as the son of a Texas dairyman and grew up to become a technologist. He spent the first fourteen years of his professional career as a software engineer, software architect, and engineering manager working in investment banking, e-commerce, and various dot-com startups. Since 2012 he has worked as an agile coach, helping both small clients and very large ones create agile ecosystems that produce greater business value, happier customers, happier engineers, and higher-quality products. Carpenter has a Bachelor's of Science in Physics from Texas A&M University.

Acknowledgments

I would like to thank everyone who provided feedback and insights during the development of this book. I especially thank Mary Beth Anderson, Dmitry Barsky, Dee Carpenter, and David Stackleather for all their time spent proofreading various early drafts.

I was extremely fortunate to find my editor, DeAnna Burghart, via a reference from Robert Galen. Her previous experience editing similar books in the agile space resulted in exceptional, insightful edits, matched only by her overall professionalism.

Deb Tremper's design skills, knowledge of InDesign, hard work, and patient willingness to help me improve my own design skills greatly improved the quality of the book. I am also indebted to Linh Thoi for her excellent ebook conversion; I was lucky to find her.

Part I
Conceptual Foundations

Forging
Change

Agile
Deployment
Models

Agile
Design
Elements

Mgmt.
Behaviors
in Scrum

Estimating
Business
Value

Progressive
Refinement
at Scale

Sprint
Alignment
Wall

User Story
Ruler

Example
Scrum
Task Boards

Definition of
Done
Examples

The worker is not the problem.
The problem is at the top! Management.

W. Edwards Deming

Triage
Guidelines

Ratcheting
Definition of
Done

Forecasting
Releases

Discerning
Genuine
Unit Testing

Terminology
Definitions

Training
Concerns

Scrum
Master
Selection

Scrum
Diagrams

Official
Scrum
Guide

Virtual
Kanban

J. Carpenter

1   *Agile Deployment Models*

Truth is often multifaceted, especially in the complex work of changing the culture of engineering organizations. Often, the more actionable, insightful facets are obscured by more politically palatable aspects with broader marketing appeal. This doesn't necessarily make the politically palatable aspects any less true; they simply don't provide a complete understanding.

1.1   STRUCTURE DRIVES GROUND-LEVEL CULTURE

At a high level there are at least two key leadership facets to agile adoption. Let's call the first *cultural leadership* and the second *structural leadership*. You will quickly recognize cultural leadership as the usual wisdom espoused in almost every management book. In contrast, structural leadership is very uncomfortable for many to discuss and therefore seldom given the emphasis and exposure it deserves.

Forging Change

Agile Deployment Models

Agile Design Elements

Mgmt. Behaviors in Scrum

Estimating Business Value

Progressive Refinement at Scale

Sprint Alignment Wall

User Story Ruler

Example Scrum Task Boards

Definition of Done Examples

1.1.1   Cultural Leadership

Cultural leadership refers to the insightful yet seldom controversial material you are likely already aware of. In many cases it is a bit too abstract to apply directly, unless something you read just happens to resonate with a specific challenge you are experiencing at that time.

- **Leadership Must Lead.** Executive management has an ethical and professional responsibility to establish clarity of organizational purpose, validate high-level mission intent, and model desired cultural values.

- **Management Books and Related Content.** Much has been written on cultural leadership:

  » Various books by Dale Carnegie

  » Various books by W. Edwards Deming

  » *Great by Choice* by Jim Collins and Morten T. Hansen

  » *Tribal Leadership: Leveraging Natural Groups to Build a Thriving Organization* by Dave Logan, John King, and Halee Fischer-Wright

  » *How Google Works* by Eric Schmidt and Jonathan Rosenberg

  » *Turn the Ship Around!* by L. David Marquet

  » *Drive: The Surprising Truth About What Motivates Us* by Daniel H. Pink

  » The Agile Manifesto (http://agilemanifesto.org/)

Cultural leadership is critical, and yet still insufficient to uproot legacy culture and replace it with a more effective culture aligned with an agile value system.

Triage
Guidelines

Ratcheting
Definition of
Done

Forecasting
Releases

Discerning
Genuine
Unit Testing

Terminology
Definitions

Training
Concerns

Scrum
Master
Selection

Scrum
Diagrams

Official
Scrum
Guide

Virtual
Kanban

J. Carpenter

1.1.2   Structural Leadership

1.1.2.1   Management Broke It, Only Management Can Fix It

In my experience, ground-level culture is driven by structure far more often than structure is driven by ground-level culture. Many large organizations spend a lot of time talking about agility, transparency, and other grand ideals; yet the experience in the trenches remains rather oppressive and fails to model any of the ideals being espoused. Appropriate structural change produces radically different outcomes, with significant cultural change within a matter of a few months if not a few weeks.

There are several process frameworks aligned with an agile value system, any of which can provide a clear road map for better aligning structure to the nature of complex engineering work. One of the more successful approaches is Scrum with Extreme Programming–style engineering craftsmanship practices. Unfortunately, Scrum is frequently distorted, abused, and then maligned by management to obscure the underlying organizational problems Scrum exposed. The difference between successful and unsuccessful change efforts can often be traced back to how much executive management understood and actively supported the effort.

Most of the problems I see in practice have very little to do with lack of cultural leadership at the executive level. Instead, I see managers without any appreciation for or understanding of empirical process control, whose negative behaviors are reinforced by preexisting structural forces established and promoted by executive management. In other words, **management broke the organization and only management can fix it.**

Forging Change

Agile Deployment Models

Agile Design Elements

Mgmt. Behaviors in Scrum

Estimating Business Value

Progressive Refinement at Scale

Sprint Alignment Wall

User Story Ruler

Example Scrum Task Boards

Definition of Done Examples

1.1.2.2   Using Structure to Mold Culture

An executive manager who wishes to radically transform organizational culture must implement structure and metrics that hold managers accountable to an agile value system. Without appropriate structure and metrics, a significant number of managers will rapidly distort intentions in an effort to protect themselves from the emotional challenges of changing their behavior.

I am a strong believer in using the carrot more than the stick to motivate behavioral change. People generally rise to your expectations; expect the best and you will usually get it. Unfortunately, the legacy behaviors and personal value systems of about a third of managers are usually too deeply entrenched for the carrot alone to work. Consequently, it is important to implement accountability mechanisms that ensure a manager's personal pain of not changing is greater than the personal pain of changing. In my experience, unless an executive sponsor is willing and able to fire people there won't be enough leverage to uproot the preexisting culture.

1.1.2.3   Executive Values Drive Structure

When using shorthand, I frequently say structure drives culture. This is not completely accurate. Working backward a bit:

Current Organizational Problems
 » *Current Organizational Structure*
  » *Poor Executive Management Decisions*
   » *Lack of Understanding in the Executive Layer*

The first chapter of Reinertsen's Flow book is broadly available as free preview content in various formats. Reinertsen distributes a PDF at http://lpd2.com /downloads/.

From this perspective, even failures in structural leadership are the result of failures in cultural leadership. If you don't believe there is a general lack of understanding in the executive layer regarding the nature of software engineering and similar complex work and how to best manage it, I challenge you to read the first chapter of *The Principles of Product Development Flow* by Donald G. Reinertsen.

Triage
Guidelines

Ratcheting
Definition of
Done

Forecasting
Releases

Discerning
Genuine
Unit Testing

Terminology
Definitions

Training
Concerns

Scrum
Master
Selection

Scrum
Diagrams

Official
Scrum
Guide

Virtual
Kanban

J. Carpenter

You may insist the value system of executive leadership is the most critical thing. I agree. But this perspective does not clearly illuminate an actionable path to change. My primary interest is in affecting large-scale organizational change. Leaders can talk about organizational change and ideals all day long and nothing meaningful seems to happen. But adopting a structure aligned to agile values and then helping executive leadership hammer the organization into that mold inevitably yields rapid, positive change and produces radically happier customers, happier engineers, and higher-quality products.

1.1.2.4   Helping Managers Accept Change

In a traditional organization doing predictable work, ground-level employees distill and refine information so those higher in the organizational chart can make an informed decision. The presumption is that those higher-ups are better placed to make a fully informed decision that accounts for all the constraints the organization faces. These managers provide value to the organization by striving to make the best decisions possible and providing clear, actionable guidance to employees working at the ground level.

In other words, managers in traditional organizations primarily obtain a sense of self-worth and importance by going to lots of meetings and making important decisions affecting those who report to them. In an agile organization the role of a manager is much different, and the source of a manager's sense of self-worth must change.

Agile organizational design presumes that the work is far too complex for managers to make an ideal decision. Instead, agile managers are responsible for ensuring that those closest to the work have the information flows they need to make an optimal decision and removing any obstacles those closest to the work cannot remove by themselves. The sense of self-worth felt by agile managers is more like that of parents who are more proud of their children's accomplishments than their own.

Forging
Change

Agile
Deployment
Models

Agile
Design
Elements

Mgmt.
Behaviors
in Scrum

Estimating
Business
Value

Progressive
Refinement
at Scale

Sprint
Alignment
Wall

User Story
Ruler

Example
Scrum
Task Boards

Definition of
Done
Examples

This is a gross oversimplification, but it highlights why some managers find their organization's agile adoption so emotionally challenging to accept. In many ways middle management has far more to lose in an agile adoption than anyone else in the organization. They have spent years becoming good at things that are no longer highly valued. This is potentially terrifying, especially when you mix in concerns about providing for family and maintaining social standing.

I can make an argument that the emotional rewards and intellectual challenges for managers are greater in agile organizations than in traditional organizations. Even though this may be true, it still requires a leap of faith to give up a comfortable, well-understood situation for a much less familiar one.

Helping managers overcome their fears during an agile adoption requires both cultural and structural leadership from executive management. We must compassionately recognize the justifiable fears involved and help people develop the courage to move past them. At the same time, we must hold management accountable for change.

I believe that using traditional management techniques for complex project work such as software engineering effectively holds employees accountable for things outside their control. I also believe that executive management's moral obligation to ensure ground-level employees are treated fairly supersedes the needs of their middle managers to feel emotionally comfortable. So while executive management should compassionately and patiently support middle managers in their transition to an agile model, any manager who isn't making a good-faith effort to come into alignment must be actively moved out of the organization.

Triage
Guidelines

Ratcheting
Definition of
Done

Forecasting
Releases

Discerning
Genuine
Unit Testing

Terminology
Definitions

Training
Concerns

Scrum
Master
Selection

Scrum
Diagrams

Official
Scrum
Guide

Virtual
Kanban

J. Carpenter

1.2   OVERVIEW OF AGILE DEPLOYMENT MODELS

There are several models for deploying agile within a large organization, and each has its strengths and weaknesses. This list uses my own terminology for the three primary models:

1.   Attractor Change Model

2.   Scrum Studio Change Model

3.   Executive Pull–Based Change Model

These models are described in greater detail over the next few pages. Each one builds on the preceding models. For example, the Attractor Change Model emphasizes helping people only in areas where they want help. This continues to be a necessary and useful part of any transformation effort, even when the later, more advanced models are also in play.

Similarly, as a company transitions to the Executive Pull–Based Change Model, there will occasionally be a need to try out new techniques and metrics before applying them more broadly. The portion of the organization that initially executed in a Scrum Studio Change Model is often the best place to conduct those experiments.

You will likely recognize some of these models by other names. For example, I often hear a Scrum studio called a pilot, a bubble, or a walled garden. I even use these names myself at times. I have used the term *Scrum studio* here out of respect for Ken Schwaber and Jeff Sutherland, as that is the term they use in *Software in 30 Days*.

George Box famously said, "All models are wrong, but some are useful." I'm sure these mental models are wrong at some level, but I have found them very useful and actionable.

Forging Change

Agile Deployment Models

Agile Design Elements

Mgmt. Behaviors in Scrum

Estimating Business Value

Progressive Refinement at Scale

Sprint Alignment Wall

User Story Ruler

Example Scrum Task Boards

Definition of Done Examples

1.2.1 Problem Statement

Before getting into the details of the various deployment models, let us first remind ourselves of the overall goal of any agile deployment and the obstacles that must be overcome.

1.2.1.1 End Goal

Delight customers with frequent, high-quality production releases meeting the customer need.

1.2.1.2 Obstacles to Change

- Reinforcing feedback loops support entrenched behavior.

- Change threatens many people's sense of self-worth, especially those in management.

- Transparency is very uncomfortable. It is human nature to avoid discomfort.

- Managing uncertainty requires accepting and embracing it.

  » **Embrace Uncertainty**: Use empirical process control to optimize outcomes.

  » **Deny Uncertainty**: Continually be frustrated by inefficiency and failure.

Triage
Guidelines

Ratcheting
Definition of
Done

Forecasting
Releases

Discerning
Genuine
Unit Testing

Terminology
Definitions

Training
Concerns

Scrum
Master
Selection

Scrum
Diagrams

Official
Scrum
Guide

Virtual
Kanban

J. Carpenter

1.2.2 Attractor Change Model

Success will naturally attract followers. Rather than attempt to radically alter people's mindset along with their organizational structure, focus improvement efforts where people are already eager to change. Promote awareness of any successful improvement efforts to help attract additional followers.

Pro:
- Builds support for additional change

- Largely avoids building resistance to change

Con:
- Seldom sufficient to fix institutionally entrenched anti-patterns

- Easily derailed as soon as anyone with managerial authority feels threatened

Forging Change

Agile Deployment Models

Agile Design Elements

Mgmt. Behaviors in Scrum

Estimating Business Value

Progressive Refinement at Scale

Sprint Alignment Wall

User Story Ruler

Example Scrum Task Boards

Definition of Done Examples

1.2.3   Scrum Studio Change Model

The concept of using an isolated bubble can be used to implement any desired reasonable agile process.

Establish a protected, volunteer-only part of the organization. All parties in this Scrum studio, including business and engineering, agree to abide by the rules of Scrum.

Pro:

- Studio will deliver excellent productivity gains

- Useful for establishing broader buy-in

- Some improvement in the legacy portions of the organization due to osmosis

Con:

- Resisters in the legacy portions of the organization are unlikely to decide to change.

- Individual contributors in legacy portions of the organization continue to suffer under unreasonable expectations.

- Legacy portions of the organization will only deliver marginal improvements.

- Legacy portions of the organization will attempt to create an illusion of change that obscures the real productivity gap.

- Isolation can be hard to achieve.

Triage
Guidelines

Ratcheting
Definition of
Done

Forecasting
Releases

Discerning
Genuine
Unit Testing

Terminology
Definitions

Training
Concerns

Scrum
Master
Selection

Scrum
Diagrams

Official
Scrum
Guide

Virtual
Kanban

J. Carpenter

1.2.4   Executive Pull–Based Change Model

This model has two key aspects:

- **Advisory:** Only provide knowledge and guidance to those who seek (pull) it.

- **Transparency:** Continually validate and publicize alignment to executive intent. Executives must ensure the pain of not changing exceeds the pain of changing. (See "1.1.2.3 Executive Values Drive Structure" on page 6.)

Pro:

- Provides a structured, actionable path to achieving real change in the entire organization

- Provides actionable guidance within each step of the Kotter change model (assumes a nonlinear view)

- Entire organization benefits from productivity gains

- Rewrites organizational DNA

Con:

- It is limited and empowered by the vision and commitment of executive leadership.

- Change is uncomfortable.

- Expect some staff turnover.

As long as the vast majority of people leaving the organization are managers who are uncomfortable serving others, is staff turnover a bad thing?

1.2.4.1   Transparency Mechanisms

To hold management accountable to an agile value system, an organization must first establish effective transparency mechanisms. Executive management must endorse and support these mechanisms and ensure everyone in the organization understands and accepts relevant changes in the role expectations.

Forging Change

Agile Deployment Models

Agile Design Elements

Mgmt. Behaviors in Scrum

Estimating Business Value

Progressive Refinement at Scale

Sprint Alignment Wall

User Story Ruler

Example Scrum Task Boards

Definition of Done Examples

Without the transparency component the Executive Pull–Based Change Model devolves to the very limited Attractor Change Model.

Great care must be taken to ensure the chosen transparency mechanisms will resist distortion; otherwise there will be a lot of ceremony and very little actual change. Without such mechanisms the Executive Pull–Based Change Model will quickly devolve into the very limited Attractor Change Model.

To make this a bit more actionable, I describe several concrete transparency mechanisms below. Each of these is a proven technique that is generally effective with enough executive management support. There is not a lot of consistency in how these techniques are named within the agile community, even though each one will likely be recognizable to a seasoned agile coach. I recommend you start with the techniques listed here, then evolve these techniques and invent new ones as you discover what works best in your own context.

1.2.4.1.1 Leadership Scrums

- Product Backlog Items (PBIs) focused on organizational changes

- Managers as Scrum Development Team members

- Executive manager as the Product Owner

- Executive Agile Coach or other appropriate choice as Scrum Master

- Usual Scrum mechanics to drive accountability

> Populating the Leadership Scrum Development Team with nothing but agile coaches and project managers will destroy the intention of holding functional and engineering managers accountable for change.

Leadership Scrum

Unlike a typical Scrum Team which develops shippable software, a Leadership Scrum is focused on creating organizational change. The Development Team roles are filled by those in the middle management layer, with the Product Backlog Items focused on improving the ecosystem. Although there is a difference in product focus and who is playing the various roles, a Leadership Scrum is structurally identical to every other Scrum Team.

Development Team: Managers

Product Owner: Executive Sponsor

Scrum Master: Initially Agile Coach

Stakeholders: Everyone else, including those who report to the managers in the Development Team.

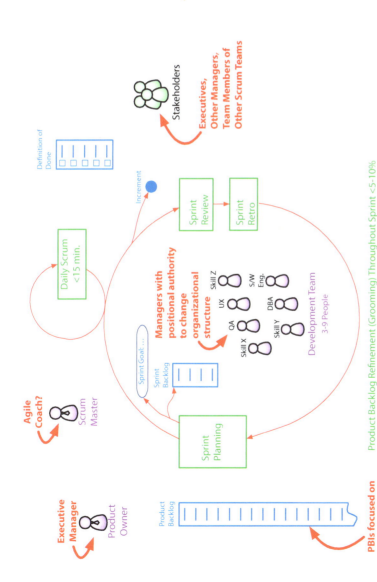

Figure 1-1 A Leadership Scrum Team organizes managers under the same execution model as any other Scrum Team. Whereas a typical Scrum Team is focused on product delivery to customers outside the organization, a Leadership Scrum Team is focused on organizational improvement. Without a mechanism to hold middle management accountable for serving product-focused Scrum Teams, middle managers will typically continue in their old behaviors. If product-focused Scrum Development Teams are fully accountable for delivery, there should be no excuse for middle management to be doing anything unrelated to improving the organization.

Forging
Change

Agile
Deployment
Models

Agile
Design
Elements

Mgmt.
Behaviors
in Scrum

Estimating
Business
Value

Progressive
Refinement
at Scale

Sprint
Alignment
Wall

User Story
Ruler

Example
Scrum
Task Boards

Definition of
Done
Examples

1.2.4.1.2   Agile Assessments

- Routine assessments of alignment to desired agile processes and desired engineering practices

- Assessment of knowledge depth in empirical process control

- Aggregated scores on each measure at each management layer

- Validated by an external expert—typically an external agile coach—whose reporting chain is independent of those being measured

- Assessment structure and results typically managed in a big spreadsheet or equivalent online tool containing clearly defined measures, assigned scores, and improvement actions, with heat maps, spider graphs and the like produced to help people see the big picture

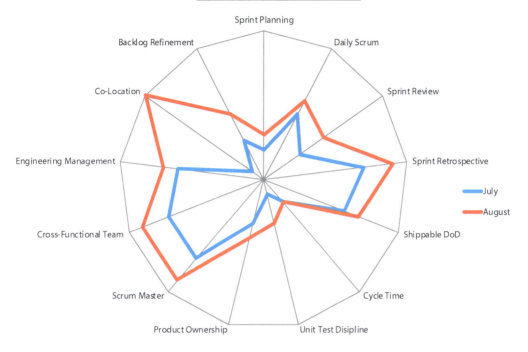

Figure 1-2 Agile assessments are used to provide transparency and trending on a variety of process and craftsmanship practices. To avoid underreporting of politically disagreeable facts, the assessments must be overseen by someone orthogonal to the reporting chain of those being assessed. To ensure any customized measures are aligned with an agile value system, the measures should be collaboratively designed by experts in agile process and engineering craftsmanship. The individual team results are typically aggregated along each measure to make the agile adoption progress of each manager or Scrum Leadership Team self-evident. The example spider graph provides a few sample measures; real-world assessments have at least twice as many.

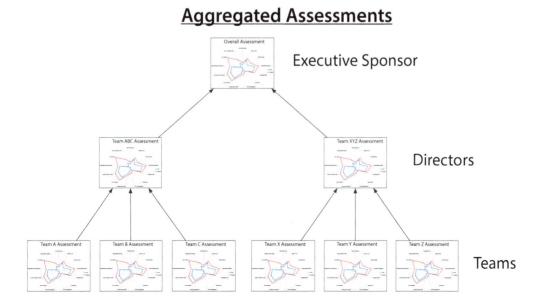

Forging
Change

Agile
Deployment
Models

Agile
Design
Elements

Mgmt.
Behaviors
in Scrum

Estimating
Business
Value

Progressive
Refinement
at Scale

Sprint
Alignment
Wall

User Story
Ruler

Example
Scrum
Task Boards

Definition of
Done
Examples

1.2.4.1.3   Obstacle Board Process

- Issue tracking of impediments discovered by agile teams that are outside the team's ability to solve without management involvement

- Less subjective escalation procedures which make it obvious when managers are failing to serve those they have the privilege to lead; for example, no obstacle should remain on a given board for more than three days before being forcibly promoted to the next-level board by a relevant Scrum Master

- Frequently inspires PBIs for the Leadership Scrum Team's Product Backlog

- Cadenced review mechanisms in which the executive manager holds management accountable for actively working and removing obstacles; frequently joined with Leadership Scrum Sprint Review

> The tactical nature and high visibility of physical boards seem to generate and maintain more social awareness than pure electronic solutions. Try to establish an effective information radiator, not an information closet.
>
> An obstacle board process typically requires the active engagement of executive management to be successful; otherwise, middle management focuses on whatever else executive management is focusing on.

Obstacle Board Detail

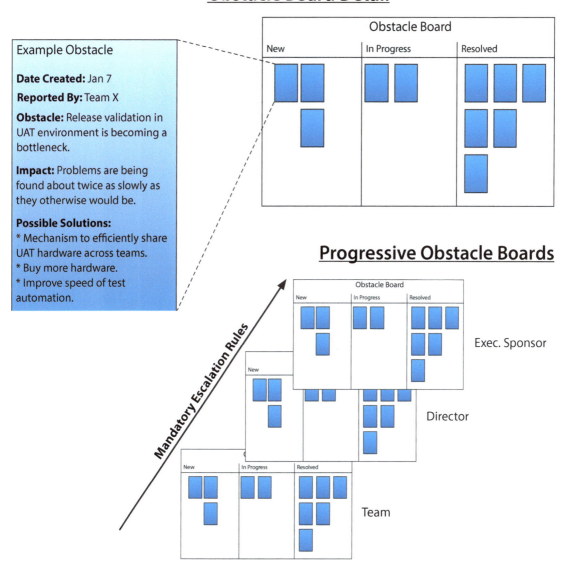

Example Obstacle

Date Created: Jan 7

Reported By: Team X

Obstacle: Release validation in UAT environment is becoming a bottleneck.

Impact: Problems are being found about twice as slowly as they otherwise would be.

Possible Solutions:
* Mechanism to efficiently share UAT hardware across teams.
* Buy more hardware.
* Improve speed of test automation.

Obstacle Board

New | In Progress | Resolved

Progressive Obstacle Boards

Mandatory Escalation Rules

Obstacle Board — New | In Progress | Resolved

Exec. Sponsor

Director

Team

Figure 1-3 An obstacle board process establishes a lightweight mechanism for tracking the life cycle of obstacles identified by the individual teams. Creating an obstacle ticket should be as effortless as possible. Even the meekest team members should be encouraged to identify obstacles and protected from any repercussions of doing so. An enforced escalation mechanism based on obstacle age or a similar measure must ensure obstacles are quickly given higher visibility when not rapidly resolved. As an example, no obstacle should remain on a given board for more than three days before being forcibly promoted to the next-level board by a relevant Scrum Master.

Forging
Change

Agile
Deployment
Models

Agile
Design
Elements

Mgmt.
Behaviors
in Scrum

Estimating
Business
Value

Progressive
Refinement
at Scale

Sprint
Alignment
Wall

User Story
Ruler

Example
Scrum
Task Boards

Definition of
Done
Examples

1.2.4.1.4   Continuous Coaching

- Continuous coaching by expert agile coaches until agile transition is largely complete

- Ensure agile coaches have a reporting structure that bypasses the group undergoing organizational change.

- Frequent communication between executive management and agile coaches

> Agile coaches can only provide transparency and guidance. Executive management must hold employees accountable for change.

Triage
Guidelines

Ratcheting
Definition of
Done

Forecasting
Releases

Discerning
Genuine
Unit Testing

Terminology
Definitions

Training
Concerns

Scrum
Master
Selection

Scrum
Diagrams

Official
Scrum
Guide

Virtual
Kanban

J. Carpenter

1.2.5   *Leadership Must Drive Change*

A mentor of mine once told me you can't push on a string. The example transparency mechanisms described for the Executive Pull–Based Change Model provide guidance for establishing strings that can create the social tension needed to motivate cultural change. Although anyone can help put the strings in place, ultimately senior management must be willing to pull on them.

When working with a new product-focused Scrum Team, I frequently encourage the team to pay careful attention to their definition of Done. I tell them it is very likely they will be challenged to prove any PBI they claim as done, and demo in the Sprint Review meets the definition of Done as well as any acceptance criteria detailed in the PBI. Shortly before the team's Sprint Review, I advise a few managers to selectively drill down and spot check more challenging line items in the definition of Done during the Sprint Review.

For example, let us assume the team's definition of Done requires automated unit tests for any new or modified code. I might meet with the chief technology officer before the Sprint Review and suggest surprising the team during the meeting with an ad hoc request that they present and demo the unit tests for one of the PBIs.

No matter how well I try to prepare the team and how transparent I am about having management hold the team accountable to the definition of Done, it isn't until managers actually follow through on holding the team accountable to the definition of Done that I see meaningful behavioral change in the teams.

This example was focused on a typical product-focused Scrum Team, but exactly the same approach can be used for Leadership Scrum Teams. Until senior management challenges the members of a Leadership Scrum Team to take items in their Sprint Backlog seriously, the mid-level

Forging
Change

Agile
Deployment
Models

Agile
Design
Elements

Mgmt.
Behaviors
in Scrum

Estimating
Business
Value

Progressive
Refinement
at Scale

Sprint
Alignment
Wall

User Story
Ruler

Example
Scrum
Task Boards

Definition of
Done
Examples

managers in the Leadership Scrum Team will continue in their old behaviors. I find it is best to initially run one-week Sprints for Leadership Scrum Teams, as they tend to be a rather stubborn group requiring a lot of reinforcement before they will change their individual focus and behaviors.

In the example above, the built-in transparency mechanisms of Scrum's Sprint Review and Definition of Done are the strings management must pull on. More broadly, a formal review of agile assessment results, an informal review of a set of obstacle boards, and routine one-on-one meetings with a senior manager's direct reports each provide an opportunity for a senior manager to demonstrate a focus on and commitment to helping the organization embrace an agile value system.

Fredric Laloux's "Reinventing Organizations" details a color based scheme for categorizing the cultural maturity and social operating model of an organization.

When attempting to implement an agile execution model, it helps to remember how critical active leadership support is in driving change. In terms of Frederic Laloux's organizational maturity models, you are probably trying to transform a predominately Orange-level organization to a Green-level organization. If you are already working in a Green organization, you probably don't need anything in part 1 of this book. If you are working in a Red or Amber organization it is highly unlikely senior management will be willing to accept much of the guidance in part 1 of this book.

1.2.6   Refining Transparency Mechanisms in a Scrum Studio

Early in the life of a Scrum studio, the combination of self-motivated volunteers and smaller scale means that Scrum's built-in transparency mechanisms are usually all that is needed. In these early stages, formal assessments create an additional, unwelcome burden for people who are already eagerly struggling to adapt to an entirely new way of working. Assuming the Sprint Retrospectives are running well, the Scrum Teams will already know their immediate problems.

Triage
Guidelines

Ratcheting
Definition of
Done

Forecasting
Releases

Discerning
Genuine
Unit Testing

Terminology
Definitions

Training
Concerns

Scrum
Master
Selection

Scrum
Diagrams

Official
Scrum
Guide

Virtual
Kanban

J. Carpenter

As a Scrum studio matures, the situation changes. The initial struggles of adopting Scrum pass, and teams gel. Many of the numerous benefits of the higher-level transparency mechanisms used in the Executive Pull–Based Change Model will now yield similar benefits within the Scrum studio. Many of the larger remaining obstacles will now be outside the control of the Scrum studio. Anything that helps clarify the various obstacles and makes them more visible inside and outside the Scrum studio increases the likelihood that the obstacles will be removed.

More importantly, the Scrum studio can be used as a petri dish for refining and adapting transparency techniques in preparation for their use in an Executive Pull–Based Change Model. Putting the high-level transparency mechanisms in place and iteratively adjusting them based on the collaborative feedback of enthusiastic Scrum studio members inevitably results in better transparency mechanisms. The resisters outside the Scrum studio will be looking for any excuse to discredit the agile adoption effort. Better to train in a safe environment, before taking action in a less forgiving environment.

1.3   REFERENCE INFORMATION

A variety of chapter-specific reference information is available on the companion website at http://forgingchange.com/fc_adm. This URL has been encoded in the QR code below for your convenience.

Forging
Change

Agile
Deployment
Models

Agile
Design
Elements

Mgmt.
Behaviors
in Scrum

Estimating
Business
Value

Progressive
Refinement
at Scale

Sprint
Alignment
Wall

User Story
Ruler

Example
Scrum
Task Boards

Definition of
Done
Examples

All models are wrong, but some are useful.

George Box

Triage
Guidelines

Ratcheting
Definition of
Done

Forecasting
Releases

Discerning
Genuine
Unit Testing

Terminology
Definitions

Training
Concerns

Scrum
Master
Selection

Scrum
Diagrams

Official
Scrum
Guide

Virtual
Kanban

J. Carpenter

2   *Agile Design Elements*

Scrum is the most dominant and clearly articulated of the agile process frameworks. As such, it is an excellent, unambiguous reference standard to facilitate a discussion around agile design elements. Even if you are trying to deploy an alternate agile process framework, understanding ideal behavior in a Scrum context will provide useful insight.

2.1   READ THE OFFICIAL SCRUM GUIDE

There are many misconceptions about Scrum. Rather than writing yet another explanation of Scrum, I believe it is best to simply point you to the official Scrum Guide. Coauthored by Ken Schwaber and Jeff Sutherland, it is effectively the reference specification for Scrum. For your convenience I have reprinted the Scrum Guide in chapter 18. Chapter 17, "Scrum Diagrams," provides diagrams and commentary that will hopefully make it easier to follow the Scrum Guide.

Scrum is an example of using structural leadership to transform culture.

Forging
Change

Agile
Deployment
Models

Agile
Design
Elements

Mgmt.
Behaviors
in Scrum

Estimating
Business
Value

Progressive
Refinement
at Scale

Sprint
Alignment
Wall

User Story
Ruler

Example
Scrum
Task Boards

Definition of
Done
Examples

2.2   AGILE PROCESS IS DESIGNED FOR COMPLEX WORK

Misinformation and half-truths about Scrum are common in marketing materials written to appeal to waterfall organizations looking for the illusion of change. If you don't know how to discern truth from marketing fiction, good people in your organization will likely be hurt and also be prevented from achieving their potential within your organization. This chapter will help you better discern truth by presenting a simple gener-

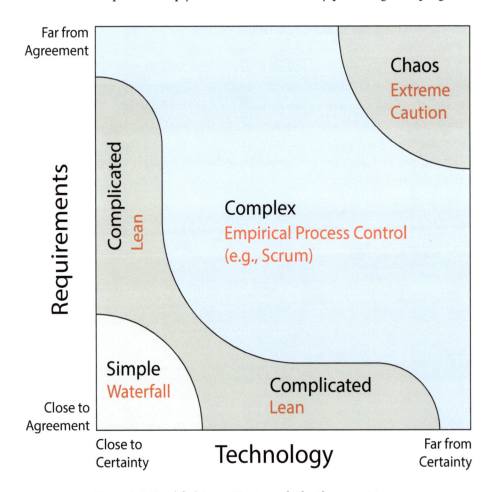

Figure 2-1 Simplified Stacey Matrix overlaid with appropriate process paradigms for each system domain.

Triage Guidelines

Ratcheting Definition of Done

Forecasting Releases

Discerning Genuine Unit Testing

Terminology Definitions

Training Concerns

Scrum Master Selection

Scrum Diagrams

Official Scrum Guide

Virtual Kanban

J. Carpenter

alized set of design elements I have found to be common to any effective agile process. As importantly, these design elements will give you a deeper understanding of Scrum and agile development fundamentals.

The clarity and insight from these design elements are very helpful, yet without the ability to relate what you are reading to at least one concrete implementation they will be very hard to understand. Scrum with a definition of Done incorporating good craftsmanship practices is the best concrete reference implementation I know of.

2.3   WATERFALL WORKS WELL FOR SIMPLE WORK

Agile process design is focused on managing work in the complex systems domain. Ralph Stacey's Stacey matrix and Dave Snowden's Cynefin model are commonly used when discussing where agile methods are most appropriate.

Many of the traditional management approaches are not necessarily wrong; they are simply inappropriate to complex project work such as software engineering. Management's obligation to serve those they lead never changes, but how servant-leadership is best performed changes based on the nature of the work.

When managing simple work, management should ensure that people have clear direction on how to best accomplish their work and that they are appreciated for a job well done. In contrast, managing complex work requires far more sophistication, highly skilled and educated knowledge workers, and a system that empowers knowledge workers with the information and autonomy needed to make optimal decisions. Most of the common problems in software engineering are a direct result of management attempting to apply an inappropriate management paradigm which is disrespectful of the nature of the work, and therefore unintentionally disrespectful of the people doing the work.

For a comparison and deeper discussion of the Stacey matrix versus the Cynefin model I recommend reading "Cynefin Framework versus Stacey Matrix versus network perspectives."[1]

Please be very mindful that "complicated" and "complex" are separate and distinct systems domains. The terms are frustratingly similar.

Forging
Change

Agile
Deployment
Models

Agile
Design
Elements

Mgmt.
Behaviors
in Scrum

Estimating
Business
Value

Progressive
Refinement
at Scale

Sprint
Alignment
Wall

User Story
Ruler

Example
Scrum
Task Boards

Definition of
Done
Examples

2.4   AGILE PROCESS DESIGN ELEMENTS

The Agile Manifesto (http://www.agilemanifesto.org/) nicely conveys the spirit of agile philosophy. Unfortunately, it isn't very actionable. "The Principles behind the Agile Manifesto" (http://www.agilemanifesto.org /principles.html) is only a little better.

After years of studying and practicing agile software engineering, I have discovered that any approach implementing the six design elements detailed in this chapter tends to be successful. Any approach which does not tends to fail in some way.

These design elements are more actionable than the Agile Manifesto and less overwhelming than Reinertsen's *Principles of Product Development Flow*. There is some overlap with David Anderson's "6 Practices for Evolutionary Design." These design elements are intended as acceptance criteria for an effective agile process, rather than as general aspirational principles. They are complementary and largely orthogonal to the principles championed by Anderson, Reinertsen, and the Agile Manifesto.

You may notice that the first three design elements come from Lean theory. These elements provide the basics of flow control, but not much else. Although flow control alone is perhaps enough to keep sufficiently simplistic manufacturing lines under control in the short run, complex efforts such as software engineering will quickly spin out of control unless the additional design elements are also accounted for.

Table 2-1 Successful agile development processes provide solutions to each of the following agile design elements. This is not a list of principles so much as a list of acceptance criteria for a successful agile process.

Design Element	Brief Description
Estimates as Estimates	**Accepting the natural variability of the work, and allowing the worker to commit to quality rather than any effort estimates.** Kanban's focus on Pull rather than Push, the agile concept of never locking both time and scope, and a focus on quality first are all interrelated facets of the same concept.
Buffer Management	**Managing the amount of work allowed to accumulate in different parts of the system.** In Kanban the concept is manifested as Work In Progress limits. A Scrum Development Team's focus on not planning more work for a Sprint than the team believes is reasonable is another form of buffer management.
Queue Prioritization	**Mechanisms for somehow deciding which work items are most important.** Scrum's force-ranked Product Backlog is a simple implementation of this concept. Kanban systems using different classes of service and associated pull rules is a more sophisticated implementation of this concept.
Fast Interpersonal Feedback Loops	**Formal and informal mechanisms for ensuring rich communication between all the individuals involved.** Scrum's formal ceremonies (Sprint Planning, Daily Scrum, Sprint Review, and Sprint Retrospective) are more formal examples of this concept. Pair programming and co-location are other examples of this concept.
Fast Technical Feedback Loops	**Technical approaches for validating system changes and detecting problems as quickly as possible.** Automated unit tests, compiler errors, static code analysis, the tooling aspects of continuous integration and continuous delivery, and automated higher-level tests are all examples of fast technical feedback loops. Many aspects of software craftsmanship are focused on establishing, maintaining, and continually striving to accelerate technical feedback loops.
External Customer Focus	**Continual focus on serving customers outside of the company.** A focus on User Stories meeting the INVEST test is one manifestation of this concept. Another example is Lean theory's focus on the global rather than the local optima.

Forging Change

Agile Deployment Models

Agile Design Elements

Mgmt. Behaviors in Scrum

Estimating Business Value

Progressive Refinement at Scale

Sprint Alignment Wall

User Story Ruler

Example Scrum Task Boards

Definition of Done Examples

2.4.1   Estimates as Estimates

2.4.1.1   Example Implementations

- Scrum's Sprint Planning event produces a "forecast" not a commitment.

- Kanban's use of *pull* rather than *push* flow mechanisms treats estimates as estimates.

2.4.1.2   Context

Reviewing the diagrams on the next few pages as you read will make this explanation easier to follow.

As Eli Goldratt explained in *Beyond the Goal*, treating estimates as commitments for complex work conflicts with the individual's desire to be seen as a reliable person. This destroys quality, morale, and forecast accuracy. I believe living in denial of this principle, rather than managing within it, is the root cause behind most of the pain and suffering in waterfall organizations.

To understand the conflict you must first accept how difficult it is to predict complex work. Common cause variation in software engineering is frequently said to be around 400 percent—that is, an incremental bit of functionality may take one-fourth of the estimated effort or up to four times the estimated effort. This remains true even with a professional software engineer who is already very familiar with the system being extended. Said another way, the probability of completion curves for complex work are very far from Gaussian and have very long, fat tails.

In contrast, other types of work are far more predictable and have much narrower probabilities of completion. In most types of work, a modest buffer in an estimate along with some contingency planning is sufficient to ensure the work will be complete within the estimate. For example, better dry cleaners, restaurants, and yard service companies typically complete their service when they say they will. The same is mostly true for reputable housing contractors and airlines, who must contend with inclement weather and other variables typical of work in the complicated systems domain.

Triage
Guidelines

Ratcheting
Definition of
Done

Forecasting
Releases

Discerning
Genuine
Unit Testing

Terminology
Definitions

Training
Concerns

Scrum
Master
Selection

Scrum
Diagrams

Official
Scrum
Guide

Virtual
Kanban

J. Carpenter

Software engineers asked to provide an effort estimate intuitively understand the tremendous variability implicit in the work. Assuming they feel safe focusing on quality and never worry the estimates will be used against them, the cumulative probability of completion for a large number of thinly sliced enhancements will become far more predictable as a whole than the probability of completion of any single slice. The cumulative probability will have a longer tail than the business might like, but the overall forecast accuracy will be far better than a waterfall organization is capable of.

On the other hand, if engineers fear the estimates will be used against them, they must give an outer estimate to protect themselves. If the work is taking less time than estimated, they will slow down, or perhaps introduce additional functionality to prolong the development activity. If the work takes longer than estimated, they must cut corners on quality and claim the work is done even when it is not. Often these behaviors are not conscious; they are simply subconscious reactions to completely unrealistic expectations imposed upon software engineers by the corporate culture and structure in which they work.

When estimates are treated as estimates, the probability of completion for each individual slice is skewed to the far right. As a result, the cumulative probability distribution is now skewed to the far right as well. In Goldratt's terms, the engineers have "kept the safety for themselves" rather than returning the safety buffer to the project.

When I explain this concept to managers in software engineering organizations, many have trouble accepting it. Those less familiar with software engineering sometimes struggle to accept that the variability is as high as it is, since it is so different from their experiences with most other work. Even more common are engineering managers who believe their engineers are already being held to very reasonable standards. Yet when I explain the concept to the engineers working under those managers, they instantly accept it and find it resonates strongly with their daily experiences.

Ensuring estimates are treated purely as estimates when working in the complex systems domain and understanding the cultural devastation wreaked by ignoring this principle is the most important concept in this book.

Forging
Change

Agile
Deployment
Models

**Agile
Design
Elements**

Mgmt.
Behaviors
in Scrum

Estimating
Business
Value

Progressive
Refinement
at Scale

Sprint
Alignment
Wall

User Story
Ruler

Example
Scrum
Task Boards

Definition of
Done
Examples

Natural Expected Time of Completion Curves=>

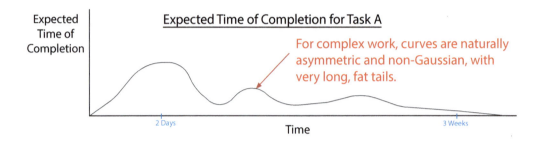

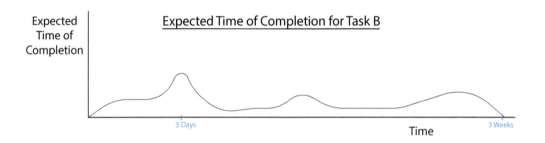

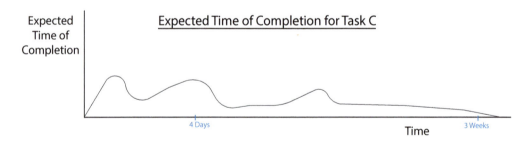

Assume each task depends upon completion
of the previous task (i.e., task seq.: A->B->C).

Figure 2-2 The natural probability of completion curves for work in the complex systems domain are highly asymmetric with very long, fat tails. The actual curve for any given complex undertaking involves too many unknowns to predict with the specificity shown here. With such long, fat tails, the time required to achieve a 50% probability of completion tends to be a small fraction of the time required to achieve a 95% probability of completion.

Triage
Guidelines

Ratcheting
Definition of
Done

Forecasting
Releases

Discerning
Genuine
Unit Testing

Terminology
Definitions

Training
Concerns

Scrum
Master
Selection

Scrum
Diagrams

Official
Scrum
Guide

Virtual
Kanban

J. Carpenter

Estimates Treated Purely as Estimates=>

Figure 2-3 Although the individual probabilities of completion are very broad, the cumulative probability distribution of a large number of dependent efforts tightens as the number of thin slices of functionality increases. It is impossible to completely remove the risk of each effort taking the outermost estimate, yet with enough thin slices the risk becomes far more manageable. All this assumes engineering effort estimates are treated purely as estimates, never commitments. As you will see in the next graph, doing otherwise destroys personal safety and artificially skews all the probabilities to the far right.

Forging
Change

Agile
Deployment
Models

Agile
Design
Elements

Mgmt.
Behaviors
in Scrum

Estimating
Business
Value

Progressive
Refinement
at Scale

Sprint
Alignment
Wall

User Story
Ruler

Example
Scrum
Task Boards

Definition of
Done
Examples

Treating estimates as commitments for complex project work is in conflict with the desire of the individual to be seen as a reliable person.

It destroys:

- Quality

- Morale/Transparency

- Forecast Accuracy

Paraphrase of Eli Goldratt, *Beyond the Goal*

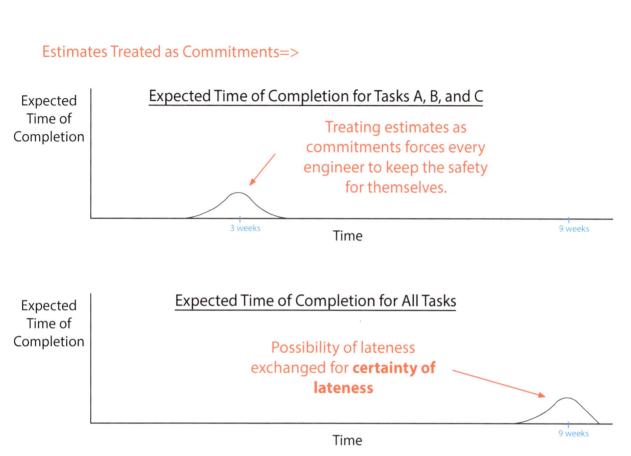

Estimates Treated as Commitments=>

Expected Time of Completion

Expected Time of Completion for Tasks A, B, and C

Treating estimates as commitments forces every engineer to keep the safety for themselves.

3 weeks — 9 weeks

Time

Expected Time of Completion

Expected Time of Completion for All Tasks

Possibility of lateness exchanged for **certainty of lateness**

9 weeks

Time

Figure 2-4 Treating effort estimates as commitments results in engineers providing engineering effort estimates skewed as far to right as they believe will be politically acceptable. If the work takes much less time than the estimate, the engineers will subconsciously invent reasons to take longer, since delivering significantly early will be perceived as being unreliable. If the work takes much longer than expected, the engineers will compromise quality, since delivering late is perceived as being more unreliable than producing low-quality product. By attempting to hold engineers accountable for something outside their control, management ensures the cumulative probability distribution is artificially skewed far to the right while destroying quality, transparency, and morale.

Forging
Change

Agile
Deployment
Models

**Agile
Design
Elements**

Mgmt.
Behaviors
in Scrum

Estimating
Business
Value

Progressive
Refinement
at Scale

Sprint
Alignment
Wall

User Story
Ruler

Example
Scrum
Task Boards

Definition of
Done
Examples

Once again, treating estimates as commitments for complex work conflicts with the individual's desire to be seen as a reliable person. Failure to accept this reality and manage within it will destroy quality, morale, and transparency as well as forecast accuracy. This is the most important concept in this entire book.

2.4.1.3 Commitments Distort Completion Curves

See figure 2-2, figure 2-3, and figure 2-4 for a graphical explanation of the devastating problems caused by treating estimates as commitments when working in the complex systems domain.

Triage
Guidelines

Ratcheting
Definition of
Done

Forecasting
Releases

Discerning
Genuine
Unit Testing

Terminology
Definitions

Training
Concerns

Scrum
Master
Selection

Scrum
Diagrams

Official
Scrum
Guide

Virtual
Kanban

J. Carpenter

2.4.1.4   Objections You May Hear

2.4.1.4.1   I Always Gave Accurate Estimates

Some engineering managers may argue that their expectations of accurate estimates are no more restrictive than those they once successfully worked within. When I drill down into this response, I inevitably find a lack of well-defined quality standards. These managers typically have a poor awareness of what good code looks like and fail to understand the long-term productivity costs of poor code quality. In defense of these managers, they have probably spent their entire career within a context that gave lip service to quality yet in practice held them accountable to dates above all else.

Interestingly, other first-level engineering managers in these same organizations are often fully aware there is a serious quality problem, yet feel powerless to fix it within a waterfall context. These more knowledgeable managers frequently report to the managers above who refuse to accept responsibility for maintaining high quality standards.

2.4.1.4.2   Aren't You Being Overly Dramatic?

Another objection is that it is a bit extreme and exaggerates the magnitude of the problem to say treating estimates as commitments "conflicts with the individual's desire to be seen as a reliable person." Think about it a little further and you will see the problem really is as severe as Goldratt says. As Daniel Pink suggests, most people find themselves in professional creative work in large measure due to intrinsic motivations.[2] As managers, our focus should be on creating ecosystems that nurture and reinforce those intrinsic motivations. That said, most people's primary concern is providing for themselves and their family. Being seen as a diligent, honorable person is critical to maintaining organizational status, career progression, and even a sense of self-worth. Creating a situation which inadvertently stacks the best interest of the company against a person's honor is guaranteed to be problematic.

Forging
Change

Agile
Deployment
Models

Agile
Design
Elements

Mgmt.
Behaviors
in Scrum

Estimating
Business
Value

Progressive
Refinement
at Scale

Sprint
Alignment
Wall

User Story
Ruler

Example
Scrum
Task Boards

Definition of
Done
Examples

2.4.1.4.3   The Business Needs Estimates as Commitments to Survive

Market conditions and other economic constraints frequently make shipping by a given date critical to a business. But a well-run agile team can manage to a fixed delivery date far better than a waterfall team by incrementally refining scope. I won't review here how agile teams deal with fixed dates. The critical thing to accept, both emotionally and analytically, is that treating estimates as commitments is not the solution to delivery date pressure.

The nature of software engineering and other complex work requires empirical process control to successfully predict completion and deliver a high-quality product. Galileo Galilei's assertion that the earth revolved around the sun was extremely hard for clergy in the Catholic Church to accept, but that didn't make it any less true. A ship captain trying to avoid being lost at sea is unlikely to be overly concerned about religious doctrine if celestial navigation yields the most precise location fix possible. It is far more practical to embrace physics and manage within it than to suffer the consequences of living in denial.

It is always possible an engineering effort will fail. Agile methods can adapt to a dynamic business environment far better than waterfall can, yet brutal economic realities can sometimes overwhelm the best efforts of any management paradigm. A well-executed agile effort will at least help us to predict failure sooner and cut our losses.

Triage
Guidelines

Ratcheting
Definition of
Done

Forecasting
Releases

Discerning
Genuine
Unit Testing

Terminology
Definitions

Training
Concerns

Scrum
Master
Selection

Scrum
Diagrams

Official
Scrum
Guide

Virtual
Kanban

J. Carpenter

2.4.1.4.4   No One Really Treats Estimates as Commitments

Anyone who has worked in a waterfall software shop for any length of time will immediately recognize that delivery schedules are almost never built from the bottom up. Even waterfall shops long ago threw away complex Gantt charts with detailed, carefully estimated work breakdown structures. Everyone knows that if you add up all the individual bottom-up estimates (read "commitments") you end up with outrageously conservative delivery forecasts.

In practice, senior management on the business side negotiates with senior management on the engineering side to decide which items in a large, force-ranked list of features are in scope and which are out of scope for a desired delivery date. The business is always pushing for more, while engineering is always pushing for less. Although this approach has serious problems, in the short term it sort of works. More experienced managers and architects get pretty good at guessing.

But quality still suffers. Inevitable slips in schedule and quality destroy trust between the business and engineering, and forecast accuracy is also destroyed. So even though more sophisticated waterfall engineering organizations don't explicitly treat estimates as commitments at the individual level, the overall behavioral and cultural forces are close enough to cause all the same problems.

2.4.1.5   Goldratt's Lecture on Estimation

If you are struggling with my explanation, I challenge you to listen to Goldratt's related lecture in the audiobook of *Beyond the Goal*. You can find it about 1 hour and 46 minutes into the recording. The audiobook includes a reference guide with Goldratt's slide deck, which will make following the lectures much easier. The CD version doesn't include the slide deck.

Forging
Change

Agile
Deployment
Models

Agile
Design
Elements

Mgmt.
Behaviors
in Scrum

Estimating
Business
Value

Progressive
Refinement
at Scale

Sprint
Alignment
Wall

User Story
Ruler

Example
Scrum
Task Boards

Definition of
Done
Examples

2.4.1.6   This Is Why Agile Forbids Locking Time and Scope

Agile literature often says you should never lock both time and scope. This is basically the same principle as treating estimates as estimates, but it doesn't get to the root of the issue as effectively as Goldratt's explanation does. Cursory discussions also fail to explain that locking time and scope for complex systems domain work will completely destroy trust and cohesion within an organization.

2.4.1.7   Beware of Those Who Obscure the Inconvenient Truth

Many people will try to soften this message because it is such an inconvenient truth. Don't believe them. The only way to protect ourselves from the uncertainty inherent in creative software engineering efforts is to always keep our code in a shippable state, and put lots of super-fast feedback loops in place to ensure we can quickly adapt to whatever comes our way.

Triage
Guidelines

Ratcheting
Definition of
Done

Forecasting
Releases

Discerning
Genuine
Unit Testing

Terminology
Definitions

Training
Concerns

Scrum
Master
Selection

Scrum
Diagrams

Official
Scrum
Guide

Virtual
Kanban

J. Carpenter

2.4.2   Buffer Management

2.4.2.1   Example Implementations

- Kanban systems use explicit work in progress (WIP) limits to manage buffers.

- The use of the Sprint Backlog coupled with the Sprint Goal as described in the official Scrum Guide is effectively a WIP limit in disguise.

- Freeway on-ramp metering schemes designed to keep the freeways operating at optimum capacity are a real-world example.

2.4.2.2   Context

It is important that we somehow limit the amount of WIP at any one time. Otherwise, we will operate the system less efficiently than we should. This is true in both complicated and complex system domains.

Lack of buffer management also causes social issues since humans don't multitask well at the individual or organizational level. For more details see Donald Reinertsen's *Principles of Product Development Flow*.[3] Even the first few chapters drive this point home. Goldratt's book *The Goal* also addresses this topic multiple times.

Forging
Change

Agile
Deployment
Models

Agile
Design
Elements

Mgmt.
Behaviors
in Scrum

Estimating
Business
Value

Progressive
Refinement
at Scale

Sprint
Alignment
Wall

User Story
Ruler

Example
Scrum
Task Boards

Definition of
Done
Examples

2.4.3   Queue Prioritization

2.4.3.1   Example Implementations

- Scrum's use of a force-ranked Product Backlog

- Kanban's use of sorted queues, classes of service, and explicit pull rules

- Various Quality of Service strategies in a network stack

2.4.3.2   Context

David Anderson has written extensively about using Kanban for software engineering efforts. From Anderson's perspective, a Virtual Kanban system is a concrete implementation of only the first four of his five key principles in his Principles of the Kanban Method.

We must somehow make hard choices between what will and won't get done. Regardless of specific queue prioritization strategy, we must also ensure that pressing things don't perpetually overwhelm more important but less pressing things.

Triage
Guidelines

Ratcheting
Definition of
Done

Forecasting
Releases

Discerning
Genuine
Unit Testing

Terminology
Definitions

Training
Concerns

Scrum
Master
Selection

Scrum
Diagrams

Official
Scrum
Guide

Virtual
Kanban

J. Carpenter

2.4.4   *Fast Interpersonal Feedback Loops*

2.4.4.1   *Example Implementations*

- Scrum's formal events/ceremonies

- Co-located teams

- Pair programing

2.4.4.2   *Context*

We must ensure optimal communication between everyone involved, leveraging process, physical configuration of team space, frequent releases, and anything else which helps us do so. Much of the formalized process of Scrum is focused on optimizing human communication.

Forging
Change

Agile
Deployment
Models

Agile
Design
Elements

Mgmt.
Behaviors
in Scrum

Estimating
Business
Value

Progressive
Refinement
at Scale

Sprint
Alignment
Wall

User Story
Ruler

Example
Scrum
Task Boards

Definition of
Done
Examples

2.4.5   Fast Technical Feedback Loops

2.4.5.1   Example Implementations

- Automated tests at every level of the test pyramid

- Keeping the code clean and readable at all times

- Continuous integration as a behavior and as a tool

- Continuous delivery as a behavior and as a tool

- Focus on unit-level versus integration-level testing

- Technical practices associated with Extreme Programming

2.4.5.2   Context

Unless we can detect problems very quickly, the system is likely to spin out of control. We are in the business of engineering. All the fancy process in the world won't save us if we fail to practice great craftsmanship.

It can be useful to think of very quick-running automated tests as a form of asynchronous communication between engineers. A software engineer who writes an effective automated unit test is communicating to anyone who later modifies the system. With a full suite of quick-running unit tests, an engineer can quickly validate whether a change is aligned with the design intent of the rest of the existing system as expressed by those who came before, including the engineer's past self.

Triage Guidelines

Ratcheting Definition of Done

Forecasting Releases

Discerning Genuine Unit Testing

Terminology Definitions

Training Concerns

Scrum Master Selection

Scrum Diagrams

Official Scrum Guide

Virtual Kanban

J. Carpenter

2.4.6   External Customer Focus

2.4.6.1   Example Implementations

- "Valuable" in the INVEST acronym guiding User Story creation

- A focus on having actual end customers in a Sprint Review

- The Scrum Guide's focus on active engagement with stakeholders

2.4.6.2   Context

There is always a danger that a team will become focused on serving its own needs rather than maintaining an external customer focus. So it's important that your process emphasizes the need to deliver actual value to the customer.

2.5   KANBAN TECHNIQUES ARE USEFUL, EVEN IF INSUFFICIENT

You will notice that Virtual Kanban systems implement the first three design elements, but none of the others. This doesn't make Virtual Kanban systems bad, simply limited in scope. (When referring to Virtual Kanban I specifically mean the visualization of discrete quantums of knowledge work on a Kanban board with explicit WIP limits, explicit exit criteria on each activity state, and pull-based flow semantics.)

A more nuanced discussion of Kanban, Virtual Kanban, and their utility and relationship to Scrum can be found in chapter 19.

Forging
Change

Agile
Deployment
Models

Agile
Design
Elements

Mgmt.
Behaviors
in Scrum

Estimating
Business
Value

Progressive
Refinement
at Scale

Sprint
Alignment
Wall

User Story
Ruler

Example
Scrum
Task Boards

Definition of
Done
Examples

2.6   TRUST YOUR GUT

At least one popular "agile" framework I am aware of includes a variety of great design elements, yet still feels like it's more about top-down waterfall-style power dynamics than truly empowering the people closest to the work. If a process framework champions agile behaviors from a marketing perspective but creates structure that does the opposite, you should be very suspicious.

The six design elements in this chapter will detect the vast majority of broken agile processes, including the troublesome "agile" process example mentioned above. That said, if your intuition tells you something is wrong you should try to figure out why.

2.7   REFERENCE INFORMATION

A variety of chapter-specific reference information is available on the companion website at http://forgingchange.com/fc_ade. This URL has been encoded in the QR code below for your convenience.

Forging
Change

Agile
Deployment
Models

Agile
Design
Elements

Mgmt.
Behaviors
in Scrum

Estimating
Business
Value

Progressive
Refinement
at Scale

Sprint
Alignment
Wall

User Story
Ruler

Example
Scrum
Task Boards

Definition of
Done
Examples

Communication comes in both words and deeds. The latter is generally the most powerful form. Nothing undermines change more than behavior by important individuals that is inconsistent with the verbal communication.

John P. Kotter, *Leading Change*

Triage
Guidelines

Ratcheting
Definition of
Done

Forecasting
Releases

Discerning
Genuine
Unit Testing

Terminology
Definitions

Training
Concerns

Scrum
Master
Selection

Scrum
Diagrams

Official
Scrum
Guide

Virtual
Kanban

J. Carpenter

3   *Management Behaviors in Scrum*

For people to follow, we as leaders must lead. The delivery problems of most software organizations are often a direct result of management applying structure and expectations that disrespect the nature of the work—and therefore the people doing the work. If there weren't a serious problem, there wouldn't be a pressing need to adopt radically different management practices.

3.1   MANAGEMENT MUST LEAD

I believe engineering management at all layers should quickly take ownership of past failures, and diligently apply themselves to mastering empirical process control. The more senior the manager within the engineering-focused areas of the business, the greater the obligation to understand empirical process control at a deeper level than others in the organization. The engineering work we do is far too complex to think process and structure are purely project management's problem.

As managers we leverage our extensive professional expertise to hire smart, capable software, test, and hardware engineers and to lead them as best we know how. If we ask these smart people to adopt agile methods and then fail to model agile behaviors ourselves, we shouldn't be surprised if they pay more attention to our actions than our words.

Fortunately, a tremendous amount has been written about empirical process control for engineering organizations. Closing any knowledge or experience gap is mostly just an issue of effort and focus, at least as long as one avoids focusing too much on the soft, fluffy, non-actionable guidance out there. I personally become suspicious anytime the guidance drifts too far away from great craftsmanship, actionable process, effective customer engagement, and driving good economic outcomes.

In an effort to be more concrete, I have listed a few ideas below for taking more aggressive action to close any knowledge or experience gaps.

Forging
Change

Agile
Deployment
Models

Agile
Design
Elements

**Mgmt.
Behaviors
in Scrum**

Estimating
Business
Value

Progressive
Refinement
at Scale

Sprint
Alignment
Wall

User Story
Ruler

Example
Scrum
Task Boards

Definition of
Done
Examples

3.1.1   *Start by Accepting Ownership*

- Accept that there is a problem and you are very likely responsible for creating it. Structure drives ground-level behavior
  and culture, and management creates the structure. If the
  current structure is producing undesirable behaviors, don't
  blame the individual contributors working within the system.

3.1.2   *Create a Stable Foundation*

- Identify a good agile coach with agile transformation experience. Pay careful attention to the first two chapters in this
  book when interviewing candidates. The following focus
  areas will catch problems with many potential candidates:

  » Create a line of questioning that provides insight
    into a candidate's beliefs around treating effort esti-
    mates as commitments. I often do this by describing
    a scenario in which the business is trying to lock
    both time and scope and asking for advice on how
    to handle the situation. Do not lead candidates to
    your preferred answer; you want to understand
    their true views on the topic.

  » Try to figure out if the candidate truly understands
    empirical process control and self-organizing teams
    or is just trying to improve the ability to micro-
    manage. Problems in this area often surface when
    discussing forecasting and the above-mentioned
    problems with trying to lock both time and scope.
    Discussions on the role of the Scrum Master also
    tend to flush out problems in this area. Problematic
    candidates will view the Scrum Master role as very
    similar to that of a traditional project manager.

Triage Guidelines

Ratcheting Definition of Done

Forecasting Releases

Discerning Genuine Unit Testing

Terminology Definitions

Training Concerns

Scrum Master Selection

Scrum Diagrams

Official Scrum Guide

Virtual Kanban

J. Carpenter

» Try to figure out how the candidate approaches an agile adoption. A better candidate will likely be aware of many of the techniques I have described in the first chapter. Just don't expect their terminology to align with my own.

» Determine if the candidate has any real-world technical experience. Drilling down into a few software engineering basics around object-oriented design, basic unit testing, or similar areas usually works well in a software engineering context. A coach who no longer writes code everyday will be a bit rusty, but the foundations and engineering mindset never quite go away. I'm not saying I would never hire a coach without an engineering background, but I find that the best coaches generally do have one.

- Ensure you have sufficient buy-in and sponsorship at a point in the organizational chart above the various traditional silos. Without such buy-in, any agile adoption efforts will likely fail in the long run.

- Be wary of the mainstream traditional management consultancies selling agile transformation services. I suspect their standard MBA toolkit and culture is very effective at restructuring traditional businesses, performing detailed up-front market analysis, and the like. When it comes to dealing with engineering organizations I seldom see good things.

- Pay enough to get good agile coaching talent. You don't always get what you pay for, but you almost never get what you don't pay for.

Forging Change

Agile Deployment Models

Agile Design Elements

Mgmt. Behaviors in Scrum

Estimating Business Value

Progressive Refinement at Scale

Sprint Alignment Wall

User Story Ruler

Example Scrum Task Boards

Definition of Done Examples

- Humility is sadly rare in the agile coaching community. Please be careful to ensure any agile coaches you hire do their best to model the servant-leadership behaviors they ask of others.

3.1.3   Actively Manage the Transformation

- Ensure an effective obstacle board mechanism is established and actively apply yourself to removing impediments the Development Teams identify.

- Read everything you can about agile development practices and theory. Spending a couple of days in a class on agile fundamentals is far from adequate for anyone in the first few levels of engineering management.

- Take advantage of any agile coaching available to you. Frequently, people trying to solve a problem already understand how to solve it in a waterfall context. In a collaborative discussion, I am often able to help them connect the dots so they see how to solve the problem in an agile context.

- Create a deep-dive study program around various agile topics. If available, an agile coach will likely be willing to help you facilitate this and provide some guidance around topics and content. But don't let the lack of an agile coach stop you.

- Consider forming Leadership Scrum Teams within the management layers, thereby learning Scrum by immersion.

3.1.4   Design a Supportive Organization

- Consistently evolve organizational design, information systems, and culture so as to empower those closest to the work.

Triage Guidelines

Ratcheting Definition of Done

Forecasting Releases

Discerning Genuine Unit Testing

Terminology Definitions

Training Concerns

Scrum Master Selection

Scrum Diagrams

Official Scrum Guide

Virtual Kanban

J. Carpenter

- Ensure the definition of Done (or equivalent) for every code base is sufficient to squeeze out known technical risks.

- Limit work in progress (WIP) while ensuring work items are designed to focus on the external customer.

- Leverage empirical process control to accurately measure and forecast system capacity, and ensure the business accepts responsibility for living within the capacity of the system.

- Encourage and help to create open and transparent communication channels between the Development Teams and the end customers.

3.1.5   Use Sprint Reviews to Drive Accountability

The Sprint Review should facilitate holding the Scrum Development Team accountable for quality—as explicitly articulated in the definition of Done—and holding the Product Owner accountable for guiding the product. This creates the social forces, structure, and accountability needed to drive improvement. While this is always true, it is particularly relevant to very young Scrum Teams. Attention to the stakeholder actions below should help ensure young Scrum Teams adopt healthy behaviors early on.

3.1.5.1   Scrum Development Team Behavior

- During the Sprint Review, ask the Scrum Team(s) to quickly prove they did some line item in the definition of Done for one of the User Stories they claim as complete and demo it. Look for something you have particular interest in or that you suspect was not completed.

Forging Change

Agile Deployment Models

Agile Design Elements

Mgmt. Behaviors in Scrum

Estimating Business Value

Progressive Refinement at Scale

Sprint Alignment Wall

User Story Ruler

Example Scrum Task Boards

Definition of Done Examples

- Look for an opportunity to celebrate how proud you are of the Scrum Development Team's commitment to quality. Young teams should worry much more about quality and creating well-sliced User Stories than about forecast accuracy. Forecast accuracy will improve naturally, as long as we don't artificially attempt to force it to do so.

- Look for an opportunity to encourage the team to maintain focus and limit WIP. Young teams frequently work concurrently on a large number of Product Backlog Items or User Stories and fail to complete any of them. This is usually because the team members are still working as individuals instead of as a team.

3.1.5.2 *Product Owner and Scrum Development Team Behavior*

- During the Sprint Review look for an opportunity to ask the Product Owner questions about general product direction and forecasting ("what by when"). Ask the type of questions which can only be answered with a well-refined Product Backlog and good empirical metrics.

3.1.5.3 *Product Owner, Scrum Master, and Stakeholder Behavior*

- Never treat estimates as commitments. Doing so will destroy forecast accuracy, transparency, and quality. If you are trying to force a given scope by a given date and not adding capacity or removing impediments, you are treating estimates as commitments.

Triage
Guidelines

Ratcheting
Definition of
Done

Forecasting
Releases

Discerning
Genuine
Unit Testing

Terminology
Definitions

Training
Concerns

Scrum
Master
Selection

Scrum
Diagrams

Official
Scrum
Guide

Virtual
Kanban

J. Carpenter

3.1.5.4   *General Considerations at the Team Level*

We must be very careful not to disempower the Scrum Team. Don't be afraid to let the Scrum Development Team fail and have nothing to show at the Sprint Review. A well-run Sprint Review followed by an effective Sprint Retrospective will quickly act as a forcing function for the Scrum Team to own their problems and radically improve. The resulting self-imposed corrective actions will be far more effective and more quickly implemented than top-down management solutions.

Be attentive to any impediments identified by the Scrum Team during their retrospectives. It is very likely the team will bump into problems only senior management can solve. Unless management takes quick and effective action, the Scrum Team will quickly lose faith in management's commitment to them and to Scrum. Establishing an obstacle board process is an excellent way for management to hold themselves accountable to serving the Scrum Development Teams.

3.2   REFERENCE INFORMATION

A variety of chapter-specific reference information is available on the companion website at http://forgingchange.com/fc_mbis. This URL has been encoded in the QR code below for your convenience.

Part II
Reference Information

Forging Change

Agile Deployment Models

Agile Design Elements

Mgmt. Behaviors in Scrum

Estimating Business Value

Progressive Refinement at Scale

Sprint Alignment Wall

User Story Ruler

Example Scrum Task Boards

Definition of Done Examples

At the end of the day, your job is to minimize output, and maximize outcome and impact.

Jeff Patton, *User Story Mapping*

Triage
Guidelines

Ratcheting
Definition of
Done

Forecasting
Releases

Discerning
Genuine
Unit Testing

Terminology
Definitions

Training
Concerns

Scrum
Master
Selection

Scrum
Diagrams

Official
Scrum
Guide

Virtual
Kanban

J. Carpenter

4   *Estimating Business Value*

When trying to decide how to best sort a Product Backlog, it can be useful to estimate the business value each Product Backlog Item (PBI) is likely to yield. There are many ways to estimate business value, each with their own strengths and weaknesses. An effective approach in one business context might be a very poor fit in another. Rather than get too deep into the details, I will provide a few basic approaches, along with references for further study.

The creation and sorting of a Product Backlog is driven by myriad complex, interrelated, competing demands—external customer needs, technical risks, technical dependencies, market opportunities, internal customer needs, Scrum Development Team maturity, and many other facets. The techniques below are focused on obtaining a rough estimate of the more obvious value of each PBI as perceived by the typical business executive outside of the engineering organization. This is a very useful lens when sorting a Product Backlog, yet far from the only one needed. For example, a persistent failure to prioritize PBIs which improve engineering excellence will destroy long-term value.

Forging Change

Agile Deployment Models

Agile Design Elements

Mgmt. Behaviors in Scrum

Estimating Business Value

Progressive Refinement at Scale

Sprint Alignment Wall

User Story Ruler

Example Scrum Task Boards

Definition of Done Examples

4.1   ESTIMATE VALUE AT THE EPIC LEVEL

It is frequently very difficult, and seldom worth the effort, to assign business value at the individual PBI level. Assigning business value at the Epic level generally yields greater benefit for a more reasonable level of estimation effort.

4.2   BASIC VALUE ESTIMATION

Consider the following simple schemes:

1. Classify each Epic as having **Low**, **Medium**, or **High** business value or equivalent numeric values of 1, 2, 3.

2. Use **Fibonacci** values (0, 1, 2, 3, 5, 8, …) just as many teams do for effort estimation. Create a **Business Value Ruler,** analogous to a User Story Ruler, to establish relative business value sizing guidelines.

These basic value estimation schemes are often all that's required to drive meaningful discussions and negotiation among a team's stakeholders. For most agile teams the overhead of more complex value estimation schemes is seldom needed. A team that routinely releases to production tends to be guided more by real-world measurement of actual value delivered and tight customer collaboration than by estimated business value. The disadvantage of these lightweight schemes is how subjective they can be. Consequently, they are potentially more error-prone than more complex schemes.

Triage Guidelines

Ratcheting Definition of Done

Forecasting Releases

Discerning Genuine Unit Testing

Terminology Definitions

Training Concerns

Scrum Master Selection

Scrum Diagrams

Official Scrum Guide

Virtual Kanban

J. Carpenter

4.3   MODERATELY COMPLEX VALUE ESTIMATION

Consider scoring each Epic on the following measures, and then combining them into a simple weighted score.

4.3.1   Business Value Measures

- **S:** Strategic Rating: 1 to 5

- **N:** Number of customers impacted by the problem being solved: 0 to infinity

- **I:** Importance of solving this problem from the customer's perspective: 1 to 5

4.3.2   Effort Measures

- **E:** Effort Estimate in Story Points: 0, 1, 2, 3, 5, 8, …

4.3.3   Computed Business Values

- Business Value = $S \times N \times I$

- Business Value Considering Effort = $(S \times N \times I)/E$

If this scheme fails to generate a useful spread of values, consider adding weight coefficients to each term.

Ideally, the number of customers impacted by the problem and the importance of the problem should be established through direct survey of a reasonable subset of customers.

Forging
Change

Agile
Deployment
Models

Agile
Design
Elements

Mgmt.
Behaviors
in Scrum

Estimating
Business
Value

Progressive
Refinement
at Scale

Sprint
Alignment
Wall

User Story
Ruler

Example
Scrum
Task Boards

Definition of
Done
Examples

4.4   ADVANCED VALUE ESTIMATION

You can spend a great deal of time devising ever-more-sophisticated techniques to estimate business value. The "Business Value Considering Effort" formula above is a flavor of Weighted Shortest Job First (WSJF) scoring, in which a cost-of-delay estimate is divided by an implementation effort estimate.

Matthew Heusser's article "Prioritize your backlog: Use Weighted Shortest Job First (WSJF) for improved ROI" provides a good, nuanced discussion of the strengths and weaknesses of various approaches to WSJF.[4]

4.5   MEASUREMENT BEATS GUESSING

Spending time and money trying to create accurate estimates of what customers want and will pay for is sometimes less productive than building and delivering something and measuring the results. An iterative approach requires only that you figure out the most important thing to build next. Even crude business value estimation techniques are often sufficient if you can iterate at low enough cost.

Investing in making frequent customer delivery practical and improving the measurement of actual customer behaviors is often more profitable than spending the same investment on improving business value estimation.

4.6   GENERAL PRODUCT OWNER GUIDANCE

For a broader view of product ownership, consider reading what Jeff Patton[5] and Roman Pichler[6] have written on the topic. I recommend starting with blog entries on their respective websites, and perhaps reading their various books as the need arises.

Triage
Guidelines

Ratcheting
Definition of
Done

Forecasting
Releases

Discerning
Genuine
Unit Testing

Terminology
Definitions

Training
Concerns

Scrum
Master
Selection

Scrum
Diagrams

Official
Scrum
Guide

Virtual
Kanban

J. Carpenter

4.7   REFERENCE INFORMATION

A variety of chapter-specific reference information is available on the companion website at http://forgingchange.com/fc_ebv. This URL has been encoded in the QR code below for your convenience.

Forging Change

Agile Deployment Models

Agile Design Elements

Mgmt. Behaviors in Scrum

Estimating Business Value

Progressive Refinement at Scale

Sprint Alignment Wall

User Story Ruler

Example Scrum Task Boards

Definition of Done Examples

Plans are worthless, but planning is everything.

Dwight D. Eisenhower

Triage
Guidelines

Ratcheting
Definition of
Done

Forecasting
Releases

Discerning
Genuine
Unit Testing

Terminology
Definitions

Training
Concerns

Scrum
Master
Selection

Scrum
Diagrams

Official
Scrum
Guide

Virtual
Kanban

J. Carpenter

5   *Progressive Refinement at Scale*

Scrum Teams are required to keep a well-refined Product Backlog, which requires accomplishing the following objectives:

- Upcoming requirements are formed into User Stories meeting the INVEST test.

- PBIs are force ranked within the Product Backlog.

- The PBI force ranking respects the authority of the Product Owner, yet still incorporates the collective wisdom of the entire team.

- PBIs are given relative effort estimates.

- The effort estimates for PBIs towards the top of the Product Backlog represent the collective wisdom of the entire team.

- PBIs at the top of the Product Backlog are well understood by the Scrum Team likely to work on them.

- There is sufficient detail within the Product Backlog to support accurate release forecasts.

- PBI sequencing accounts for unavoidable dependencies.

Just because Scrum Development Teams treat quality as a commitment and treat effort estimates purely as estimates doesn't mean planning is unimportant. On the contrary, people new to agile environments frequently discover that the level of planning detail Scrum Teams achieve far exceeds what they have experienced in waterfall.

Forging
Change

Agile
Deployment
Models

Agile
Design
Elements

Mgmt.
Behaviors
in Scrum

Estimating
Business
Value

Progressive
Refinement
at Scale

Sprint
Alignment
Wall

User Story
Ruler

Example
Scrum
Task Boards

Definition of
Done
Examples

5.1 REFINEMENT IN A SINGLE SCRUM TEAM

Even with a single Scrum Team you must achieve the refinement goals.
At a smaller scale the complexity of formalized, multistaged refinement
meetings described later is overkill.

If you are just starting a single, isolated Scrum Team, I recommend
setting up formal cadenced refinement meetings. Doing so tends to drive
team accountability and focus regarding refinement. To help the teams
avoid the typical pitfalls of ineffective backlog refinement meetings I
recommend reading a few of the articles linked in the chapter-specific
reference information. Don't panic if the refinement process evolves, so
long as the refinement objectives are still being achieved.

Triage Guidelines

Ratcheting Definition of Done

Forecasting Releases

Discerning Genuine Unit Testing

Terminology Definitions

Training Concerns

Scrum Master Selection

Scrum Diagrams

Official Scrum Guide

Virtual Kanban

J. Carpenter

5.2   REFINEMENT AT SCALE

With many Scrum Teams collaborating on a single product, coordination becomes more challenging. There are simply too many people involved for refinement to happen organically, yet the refinement objectives must still be achieved. The good news is that the process of refinement at scale is conceptually the same as in a single Scrum Team—it's just more formalized.

In a single Scrum Team, whichever team members know the most about a given User Story typically take a first pass at refining it before bringing it to the entire team for broader discussion. Since each team member's expertise differs in area and depth, the people involved in initially refining a User Story will vary based on which User Stories are involved.

With many Scrum Teams involved, each team typically sends delegates to do the initial refinement work. Delegates who need help from a subject matter expert to refine a User Story are responsible for engaging and collaborating with the subject matter expert.

5.2.1   Avoid Formal Positional Authority Differences

Scrum is very careful to ensure all members of a Scrum Development Team are equal from a positional authority perspective. I find this is very good advice.

There will certainly be differences in pay grades and experience. Furthermore, a Scrum Development Team will typically give more weight to the advice of any team member with relevant subject matter expertise. For example, if a team is discussing database table structure choices, the Scrum Development Team will typically take careful guidance from any DBAs on the team. The same goes for any area of expertise, whether it be UX, test, or anything else.

Forging Change

Agile Deployment Models

Agile Design Elements

Mgmt. Behaviors in Scrum

Estimating Business Value

Progressive Refinement at Scale

Sprint Alignment Wall

User Story Ruler

Example Scrum Task Boards

Definition of Done Examples

The key thing to realize is that Scrum still holds all Scrum Development Team members equally accountable for creating a shippable Increment meeting the definition of Done. Although every developer has focus areas which naturally result in taking greater ownership of certain aspects of the effort, in the end they are all collectively held accountable as team members. Every time I have seen formal architect roles established, their pedestal continues to grow over time until eventually it shuts down the collaborative culture Scrum is trying to nurture.

Therefore, please treat refinement delegate roles as a responsibility or role a Scrum Team member is filling in service of the team, not as anything implying formal positional authority.

5.2.2   *Minimize Formalized Cross-Team Coordination Demands*

With great craftsmanship, far less cross-team coordination is necessary. With continued focus on clean software architectures, continuous integration, test-driven development, and similar practices, it is likely little in the way of formalized cross-team Product Backlog refinement efforts will be needed.

Formalized cross-team Product Backlog refinement and great craftsmanship practices each help compensate for deficiencies in the other. Although the formalized technique shown here is sometimes needed, it should be used as sparingly as possible. Continually improving craftsmanship so as to minimize the need for heavy cross-team refinement process is the best long-term strategy.

Triage
Guidelines

Ratcheting
Definition of
Done

Forecasting
Releases

Discerning
Genuine
Unit Testing

Terminology
Definitions

Training
Concerns

Scrum
Master
Selection

Scrum
Diagrams

Official
Scrum
Guide

Virtual
Kanban

J. Carpenter

5.2.3   Alternative Confidence-Level Labels

The nomenclature used in the diagrams and tables below expresses confidence levels in terms of 20%, 50%, 80%, and 100% enumerated confidence buckets. These are somewhat arbitrarily named. Any of the following schemes would work just as well:

The scaled refinement process shown is only one possible actionable technique. Feel free to adapt or replace as necessary.

- Low, medium, high, very high

- Coarse-grained, medium-grained, fine-grained, extremely fine-grained

- Rough, moderate, smooth, very smooth

The use of percentage-based buckets runs the risk of implying a greater level of precision than exists. Advantages include brevity, language independence, and an obvious relative relationship between levels.

5.2.4   Refinement Structure Diagrams

With the above context in mind, I believe you will find the diagrams and tables below self-explanatory. Hopefully there is enough concrete detail here to make it clear which additional refinement-related roles and meetings need to be established above those needed for a single Scrum Team. I recommend you start with the approach described and then evolve over time as you learn what works best in your specific context.

Although not shown as such in the diagrams, the multistage refinement process shown could easily be visualized on an electronic or physical Virtual Kanban board. Whether or not doing so makes sense will depend on your own context.

Please also review chapter 6, "Sprint Alignment Wall." Simply skimming figure 6-1 and figure 6-2 should provide enough context to better understand how progressive refinement activities can be coordinated.

Forging Change

Agile Deployment Models

Agile Design Elements

Mgmt. Behaviors in Scrum

Estimating Business Value

Progressive Refinement at Scale

Sprint Alignment Wall

User Story Ruler

Example Scrum Task Boards

Definition of Done Examples

5.3   REFERENCE INFORMATION

A variety of chapter-specific reference information is available on the companion website at http://forgingchange.com/fc_pras. This URL has been encoded in the QR code below for your convenience.

Multiple Scrum Development

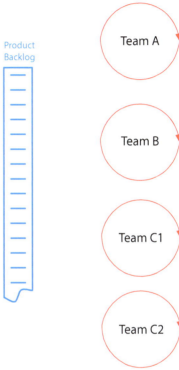

Combined Increment Must Be Shippable

The combined Increment must be shippable! It is hard to overemphasize how critical this is, and the related implications for requirements slicing and cross-team alignment.

Focus on Cross-Team Refinement

I am intentionally avoiding getting too deep into the details of how the Scrum Teams are organized. For the purpose of this chapter we will assume multiple teams whose work must be coordinated at the Product Backlog level. How to perform this coordination is largely independent of other aspects of scaling.

Reasonable Scaling Frameworks

Reasonable scaling strategies include:

- Scrum.org Nexus (http://scrum.org/nexus) and

- LeSS (https://less.works).

Nexus and LeSS are very similar, and largely codify behaviors and structures that mature large-scale Scrum Teams have been employing for years.

Figure 5-1 The use of a more formalized structure for Product Backlog refinement presumes there are too many people involved for a less formal approach to be efficient. As with any technique, the details should be adjusted as teams learn what works best in their particular context.

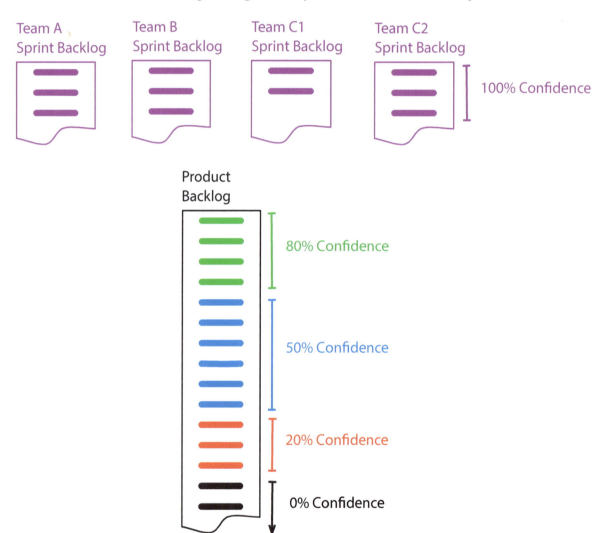

Figure 5-2 To better support coordination across Scrum Teams working on a single product, discrete levels of progressive refinement are defined. The discrete levels are not designed to imply a given level of precision so much as to signal whether enough clarity exists to justify consuming the time of a progressively larger group of Scrum Team members in additional refinement activities.

Refinement Delegation Groups

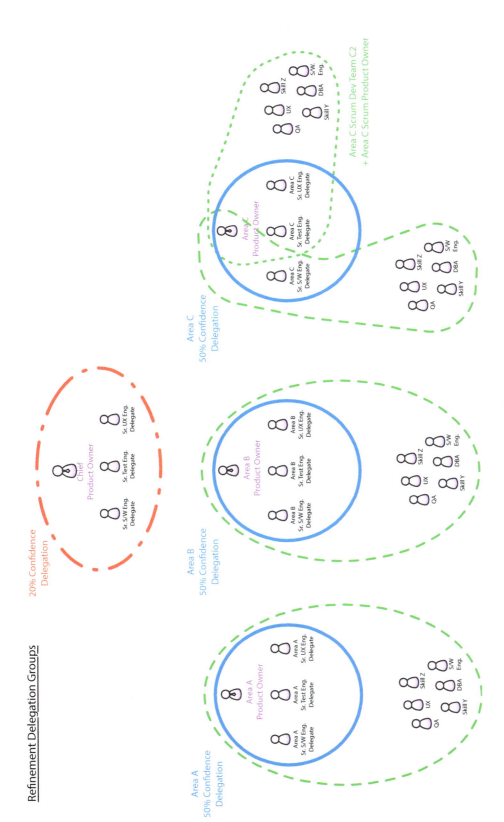

Figure 5-3 To help ensure clarity, various groups of refinement delegates are defined. The members of each group take responsibility for progressively refining PBIs in their group's area of responsibility from one discrete refinement confidence level to the next. When members of a refinement delegation group require additional expertise to refine a PBI, the relevant group is responsible for involving whoever they believe is necessary to complete the refinement.

Product Backlog with Refinement Details

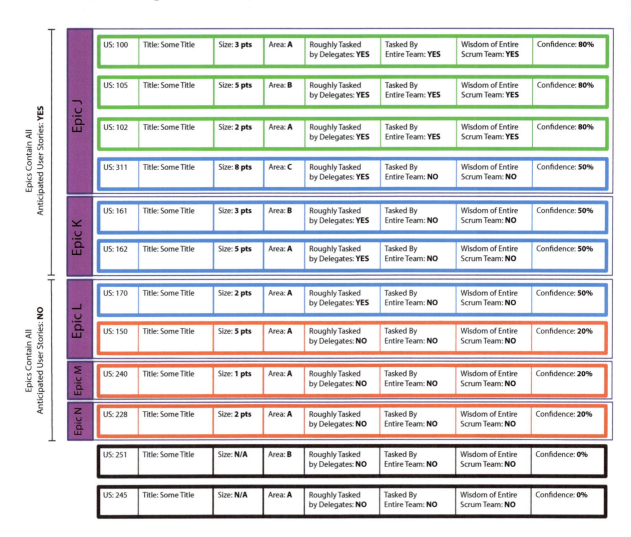

Figure 5-4 This Product Backlog diagram provides one perspective on a Product Backlog with progressively refined PBIs using the scaled refinement technique detailed in this chapter. Figure 5-5 and table 5-1 (on the next two pages) provide additional representations of the same underlying model of a Product Backlog with discrete refinement levels.

Triage Guidelines

Ratcheting Definition of Done

Forecasting Releases

Discerning Genuine Unit Testing

Terminology Definitions

Training Concerns

Scrum Master Selection

Scrum Diagrams

Official Scrum Guide

Virtual Kanban

J. Carpenter

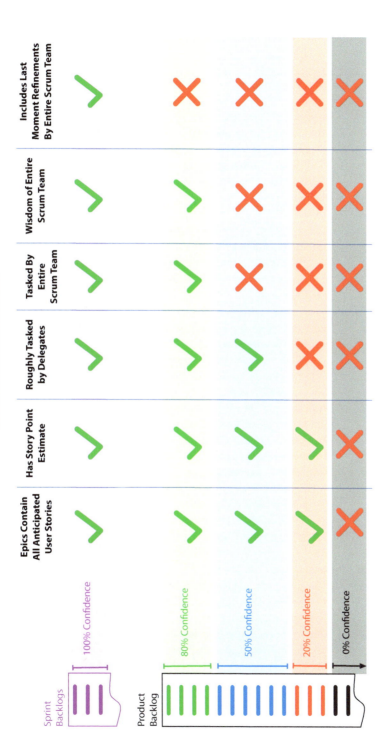

Figure 5-5 This diagram provides a tabular view of a Product Backlog with progressively refined PBIs using the scaled refinement technique detailed in this chapter. The previous diagram (figure 5-4) and table 5-1 on the next page provide additional representations of the same underlying model of a Product Backlog with discrete refinement levels.

Table 5-1 Examples of explicit refinement confidence-level guidelines are shown below. These are intended purely as a starting point. Please adapt as needed.

Confidence Level	User Story Expectations
100%	• Assigned to Current Sprint • Last moment refinements by entire team • All tasks represent less than ½ day of effort each • Placed onto Sprint task board • Every outcome from the 80% confidence level has been reviewed and revised by the entire team
80%	• Agreement within the team as to what the User Story entails • Scheduled against a Sprint • Story Points reassessed by the entire team • Every outcome from the 50% confidence level has been reviewed and revised by the entire team
50%	• Story Description is in Connextra format or equivalent • Acceptance Criteria are well written • Story Points reassessed by delegates • Tasks roughly defined • Task Hours estimated • Relevant Epic has a sufficiently complete set of User Stories • Story Description is well connected with that of the associated Epic • Acceptance Criteria are well connected with that of the associated Epic • Any end-to-end validation or other efforts required to complete the associated Epic have been accounted for by this User Story or one of its siblings within the Epic • Every outcome from the 20% confidence level has been reviewed and revised by the relevant refinement delegates • The 50% Confidence Delegation is confident the User Story is ready for review by the relevant Scrum Team(s)
20%	• Relevant Epic has been created • Relevant Epic contains at least one User Story for each related subsystem • Relevant Epic Description is in Connextra format or equivalent • Relevant Epic Acceptance Criteria are well written • Story Points assigned by delegates • Relevant subset of Epic Acceptance Criteria copied into User Story • The 20% Confidence Delegation is confident the User Story is ready for the 50% Confidence Delegation
0%	• User Story exists in Product Backlog

Table 5-2 Example refinement meeting guidelines are shown below. These are intended purely as a starting point. Please adapt as needed.

Meeting	Who	What	Suggested Cadence
80% Refinement	**_Required_** • Entire Scrum Development Team • Area Product Owner **_As Needed_** • 50% delegates from other Scrum Development Teams • Scrum Master for relevant Scrum Development Team	Refine relevant **50%** confidence User Stories **to 80%** confidence.	1–2 hours per week **per Scrum Development Team** **Every Scrum Development Team will have their own meeting.** _Adjust frequency and duration as required to achieve refinement objectives._
50% Refinement	**_Required_** • 50% delegates for relevant product area (includes PO) **_As Needed_** • Selected 80% delegates • Scrum Master for relevant product area	Refine relevant **20%** confidence User Stories **to 50%** confidence.	1–2 hours per week **per product area** **Every product area will have its own meeting.** _Adjust frequency and duration as required to achieve refinement objectives._
20% Refinement	**_Required_** • 20% delegates for relevant product (includes Chief PO) **_As Needed_** • Selected 80% delegates • Scrum Master	Refine relevant **0%** confidence User Stories **to 20%** confidence.	1–2 hours per week **per product** _Adjust frequency and duration as required to achieve refinement objectives._

Forging
Change

Agile
Deployment
Models

Agile
Design
Elements

Mgmt.
Behaviors
in Scrum

Estimating
Business
Value

Progressive
Refinement
at Scale

Sprint
Alignment
Wall

User Story
Ruler

Example
Scrum
Task Boards

Definition of
Done
Examples

We don't need an accurate document, we need a shared understanding.

Jeff Patton

Triage
Guidelines

Ratcheting
Definition of
Done

Forecasting
Releases

Discerning
Genuine
Unit Testing

Terminology
Definitions

Training
Concerns

Scrum
Master
Selection

Scrum
Diagrams

Official
Scrum
Guide

Virtual
Kanban

J. Carpenter

6   *Sprint Alignment Wall*

Coordinating work across multiple Scrum Teams is frequently challenging. Collaboratively thinking through what each team expects to work on for a few Sprints into the future helps identify dependencies and timing issues. A Sprint Alignment Wall is a simple and effective technique for facilitating this collaborative planning process.

Diagrams of two example Sprint Alignment Walls are provided on the next few pages, followed by a list of considerations to keep in mind when creating your own version.

Forging
Change

Agile
Deployment
Models

Agile
Design
Elements

Mgmt.
Behaviors
in Scrum

Estimating
Business
Value

Progressive
Refinement
at Scale

**Sprint
Alignment
Wall**

User Story
Ruler

Example
Scrum
Task Boards

Definition of
Done
Examples

6.1   BASIC SPRINT ALIGNMENT WALL

As you can see, the basic Sprint Alignment Wall is nothing other than a grid showing which User Stories each Scrum Development Team is likely to be working on for a few Sprints into the future. To start, this basic design may be all that is needed.

Although not shown here, many teams also include a column for their current Sprint.

Triage Guidelines
Ratcheting Definition of Done
Forecasting Releases
Discerning Genuine Unit Testing
Terminology Definitions
Training Concerns
Scrum Master Selection
Scrum Diagrams
Official Scrum Guide
Virtual Kanban

J. Carpenter

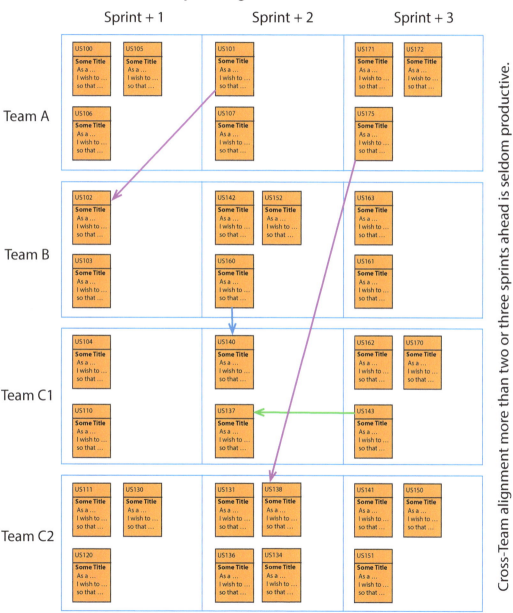

Figure 6-1 Multiple Scrum Teams working against a single Product Backlog frequently use a Sprint Alignment Wall to organize and communicate which teams are most likely to work on given PBIs. The content of a Sprint Alignment Wall should be seen as a continually evolving tentative forecast, never a commitment.

Forging
Change

Agile
Deployment
Models

Agile
Design
Elements

Mgmt.
Behaviors
in Scrum

Estimating
Business
Value

Progressive
Refinement
at Scale

Sprint
Alignment
Wall

User Story
Ruler

Example
Scrum
Task Boards

Definition of
Done
Examples

6.2 ENHANCED SPRINT ALIGNMENT WALL

After some time, teams frequently evolve more sophisticated notations to visualize additional details. The enhanced example shown here visualizes the following:

- Dependencies across Teams

- Dependencies across Time

- Epic-to-User Story Relationships

- Refinement Confidence Levels

- Release Scoping

Please don't feel constrained by this example. There are many ways to visualize this information. This is just one information radiator design out of many. With a little creativity you may soon invent an even better approach.

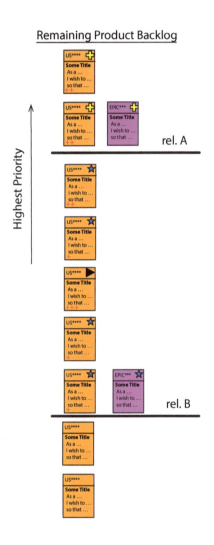

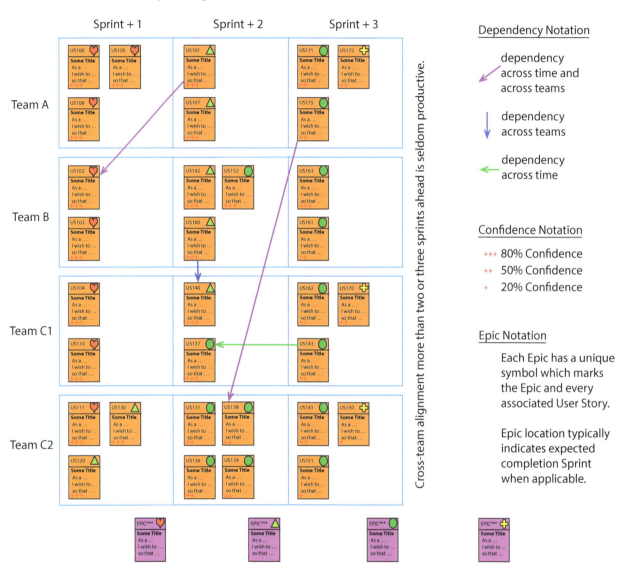

Figure 6-2 More elaborate Sprint Alignment Walls are sometimes useful. In addition to team affinity and PBI dependencies, this design includes Epic-to-PBI relationships, refinement confidence levels, the remaining Product Backlog (facing page), and release scope.

Forging Change

Agile Deployment Models

Agile Design Elements

Mgmt. Behaviors in Scrum

Estimating Business Value

Progressive Refinement at Scale

Sprint Alignment Wall

User Story Ruler

Example Scrum Task Boards

Definition of Done Examples

6.3   IMPORTANT CONSIDERATIONS

- The combined Increment must always be shippable at the end of every Sprint.

- Try to align all Product Backlog Items (PBIs) in the same Epic.

  » Overall system-level throughput and productivity is generally best when the combined Scrum Teams are focused on a very limited number of things. It is not important how productive an individual Scrum Development Team or Scrum Developer is in comparison to the throughput, productivity, and responsiveness of the combined Scrum Development Teams (see: Eli Goldratt's Theory of Constraints).

- Cross-team alignment more than two or three Sprints into the future is seldom productive. The future simply isn't that stable.

- The Sprint Alignment Wall should be viewed as a fluid, tentative forecast, which is constantly revised during refinement by self-organizing Scrum Development Teams.

Triage
Guidelines

Ratcheting
Definition of
Done

Forecasting
Releases

Discerning
Genuine
Unit Testing

Terminology
Definitions

Training
Concerns

Scrum
Master
Selection

Scrum
Diagrams

Official
Scrum
Guide

Virtual
Kanban

J. Carpenter

- The existence of a large number of dependencies between PBIs is usually a cause for concern, as it often indicates a deeper underlying problem. Here are a few of the usual suspects:

    » Poor team composition, resulting in teams which are not sufficiently independent

    » A system architecture which fails to easily absorb the most common changes required of it

    » Lack of collaborative code ownership, which is frustrated by insufficient continuous integration, unit testing, and similar craftsmanship practices needed to support it

Forging Change

Agile Deployment Models

Agile Design Elements

Mgmt. Behaviors in Scrum

Estimating Business Value

Progressive Refinement at Scale

Sprint Alignment Wall

User Story Ruler

Example Scrum Task Boards

Definition of Done Examples

6.4   CONSTRUCTION HINTS

- Don't overthink construction. Any of the following approaches will work:

    » Vinyl chart tape on a big whiteboard

    » Poster-sized self-stick paper with smaller User Story cards on top

    » Taping User Story cards to a huge wall or window

- Big visible physical Sprint Alignment Walls are often more successful than electronic versions alone. Think information radiator, not information closet.

- Tools like Rally can be used to easily print index card–sized Story Cards for a group of PBIs.

- If you use self-adhesive notes, look for the ones with extra sticky adhesive. They work much better than ordinary office notes.

- Vinyl chart tape is fantastic for delineating swim lanes and columns on whiteboards. Few office supply stores have it, but it is easy to find online.

- Consider buying a carpenter's level, tape measure, square, and large ruler. These are seldom found in the supply cabinet of today's corporate office yet very helpful when building information radiators.

Triage
Guidelines

Ratcheting
Definition of
Done

Forecasting
Releases

Discerning
Genuine
Unit Testing

Terminology
Definitions

Training
Concerns

Scrum
Master
Selection

Scrum
Diagrams

Official
Scrum
Guide

Virtual
Kanban

J. Carpenter

6.5   REFERENCE INFORMATION

A variety of chapter-specific reference information is available on the companion website at http://forgingchange.com/fc_saw. This URL has been encoded in the QR code below for your convenience.

Forging
Change

Agile
Deployment
Models

Agile
Design
Elements

Mgmt.
Behaviors
In Scrum

Estimating
Business
Value

Progressive
Refinement
at Scale

Sprint
Alignment
Wall

User
Story
Ruler

Example
Scrum
Task Boards

Definition of
Done
Examples

In theory, there is no difference between theory and practice.

But in practice, there is.

Yogi Berra

Triage
Guidelines

Ratcheting
Definition of
Done

Forecasting
Releases

Discerning
Genuine
Unit Testing

Terminology
Definitions

Training
Concerns

Scrum
Master
Selection

Scrum
Diagrams

Official
Scrum
Guide

Virtual
Kanban

J. Carpenter

7   *User Story Ruler*

Most Scrum Teams use a relative sizing measure to better quantify how much effort will be required to implement a Product Backlog Item (PBI). One or more PBIs are chosen as a reference and every other PBI is sized relative to the reference PBIs.

Most Scrum Teams implement a story point measurement standard, with size increments based on the Fibonacci sequence. The valid size sequence is therefore 0, 1, 2, 3, 5, 8, 13, 21, 34, 55, …

If everyone involved agrees upon a common reference standard for how much effort a given story point size implies, it becomes straightforward to easily plan and forecast using historical data. A User Story Ruler is a popular, effective technique to establish a common reference standard.

Given this context, the User Story diagram below is hopefully quite self-explanatory.

Forging
Change

Agile
Deployment
Models

Agile
Design
Elements

Mgmt.
Behaviors
In Scrum

Estimating
Business
Value

Progressive
Refinement
at Scale

Sprint
Alignment
Wall

**User
Story
Ruler**

Example
Scrum
Task Boards

Definition of
Done
Examples

Reference User Story Ruler

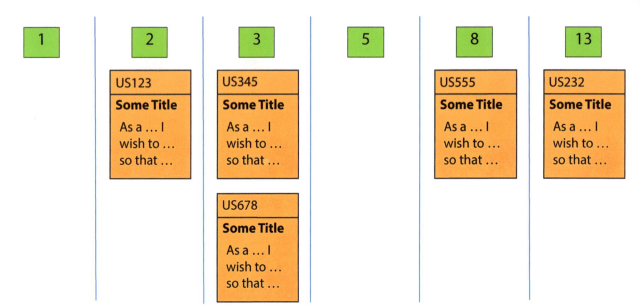

Reference User Stories serve to establish the relative story point scale.
All other User Stories are ranked relative to these reference User Stories.

Reference Story Section Criteria

- Well understood by most everyone on the Scrum Team(s).
- Math works. (For example, an 8 feels like four times more effort than a 2.)
- 13 points or smaller. (Stay within high-fidelity range.)
- Well written: good description, detailed acceptance criteria, etc.

Story Points vs. Time Estimates

- Story points focus on details of work without worrying about productivity differences between junior and senior team members.
- Story point estimation is faster than hour estimation and at least as accurate.
- Story point estimation helps avoid the temptation to treat estimates as commitments.
- Forecasting based on a past history of completed points per Sprint (velocity) provides a better quantitative basis for forecasts than time-based estimation.
- Story point estimation helps avoid giving a false sense of estimation accuracy.

Figure 7-1 A Reference User Story Ruler is a popular technique for establishing a common story point scale.

Triage
Guidelines

Ratcheting
Definition of
Done

Forecasting
Releases

Discerning
Genuine
Unit Testing

Terminology
Definitions

Training
Concerns

Scrum
Master
Selection

Scrum
Diagrams

Official
Scrum
Guide

Virtual
Kanban

J. Carpenter

7.1   SIZING TECHNIQUES

Most teams perform the estimation of User Story effort using some combination of affinity estimation and poker planning. Affinity estimation provides a very quick, low-fidelity estimate, whereas poker planning provides a higher-fidelity estimate at the cost of additional discussion time. A quick online search on "affinity estimation agile" and "poker planning agile" will find many articles on how to perform each technique.

In my experience, affinity estimation requires around a minute per PBI. A well-run poker planning session typically takes around five to ten minutes per PBI. If poker planning is taking a lot longer than this, something is likely wrong.

Affinity estimation and poker planning are not mutually exclusive. Even during poker planning sessions, I frequently group estimated PBIs by size on a nearby whiteboard. If a User Story Ruler is handy, one can place the PBIs on the ruler. Just make sure reference User Stories don't get confused with the estimated PBIs.

Treating estimates as commitments for complex project work is in conflict with the desire of the individual to be seen as a reliable person.

It destroys:

- Quality

- Morale/Transparency

- Forecast Accuracy

Paraphrase of Eli Goldratt, *Beyond the Goal*

Forging Change

Agile Deployment Models

Agile Design Elements

Mgmt. Behaviors In Scrum

Estimating Business Value

Progressive Refinement at Scale

Sprint Alignment Wall

User Story Ruler

Example Scrum Task Boards

Definition of Done Examples

7.2   WARNING: TREAT EFFORT ESTIMATES PURELY AS ESTIMATES

Software engineering involves a great deal of complexity and variability. Any given PBI may take far more or far less calendar time to complete than anyone anticipated. As soon as you start treating estimates as commitments it devastates quality, morale, transparency, and forecast accuracy. Competitively comparing story point velocity across teams will have similar effects.

7.3   EPIC BUSINESS VALUE RULER

The technique of visualizing a scale of relative sizing using concrete reference items can be applied in a variety of contexts. Creating a ruler for Epic business value is likely to be helpful when using one of the more basic schemes detailed in chapter 4, "Estimating Business Value."

7.4   REFERENCE INFORMATION

A variety of chapter-specific reference information is available on the companion website at http://forgingchange.com/fc_usr. This URL has been encoded in the QR code below for your convenience.

Forging
Change

Agile
Deployment
Models

Agile
Design
Elements

Mgmt.
Behaviors
in Scrum

Estimating
Business
Value

Progressive
Refinement
at Scale

Sprint
Alignment
Wall

User Story
Ruler

**Example
Scrum
Task Boards**

Definition of
Done
Examples

My definition of an expert in any field is a person who knows enough about what's really going on to be scared.

P. J. Plauger

Triage
Guidelines

Ratcheting
Definition of
Done

Forecasting
Releases

Discerning
Genuine
Unit Testing

Terminology
Definitions

Training
Concerns

Scrum
Master
Selection

Scrum
Diagrams

Official
Scrum
Guide

Virtual
KanBan

J. Carpenter

8   Example Scrum Task Boards

Scrum Teams typically use Scrum task boards to track the progress of the Sprint. The classic example task board seen in figure 8-1 has an elegant simplicity and flexibility which usually works better in practice than more elaborate choices. I have also given a few general recommendations based on my experience.

8.1   RECOMMENDATIONS

- Use a physical task board for co-located teams.

- Use an electronic ticket system for managing Product Backlog Items (PBIs) in the Product Backlog, and for committing source code changes against.

- Be aware that many large-scale "Agile" electronic ticket systems are catering to corporate clients who still have waterfall mindsets. These systems are usually good at organizing PBIs, yet seldom effective at forecasts based on empirical data. Don't hesitate to use spreadsheets when you need them.

- The official Scrum Guide intentionally avoids prescribing how the Scrum Development Team should track their work within the Sprint. The Scrum Development Team should be free to use whatever works best, and to evolve their techniques as they see fit.

Forging Change

Agile Deployment Models

Agile Design Elements

Mgmt. Behaviors in Scrum

Estimating Business Value

Progressive Refinement at Scale

Sprint Alignment Wall

User Story Ruler

Example Scrum Task Boards

Definition of Done Examples

8.2　REFERENCE INFORMATION

A variety of chapter-specific reference information is available on the companion website at http://forgingchange.com/fc_estb. This URL has been encoded in the QR code below for your convenience.

Sprint Task Board

Figure 8-1 This simple, classic task board design has a swim lane for each PBI and positional notation for each task state. It intentionally avoids highlighting different types of tasks, to encourage group accountability for the work. Since the design is little other than a large information radiator for an organized task list, it easily absorbs tremendous variation in task breakdown between different PBIs. Common modifications include adding WIP limits and reserving one of the swim lanes as a fire lane. Most Scrum Teams write task effort estimates on each task and plot the results on a Sprint burn-down chart. This design can easily be rotated 90 degrees to fit vertical white boards.

Triage Guidelines
Ratcheting Definition of Done
Forecasting Releases
Discerning Genuine Unit Testing
Terminology Definitions
Training Concerns
Scrum Master Selection
Scrum Diagrams
Official Scrum Guide
Virtual Kanban

J. Carpenter

Forging
Change

Agile
Deployment
Models

Agile
Design
Elements

Mgmt.
Behaviors
in Scrum

Estimating
Business
Value

Progressive
Refinement
at Scale

Sprint
Alignment
Wall

User Story
Ruler

Example
Scrum
Task Boards

Definition of
Done
Examples

The problem with quick and dirty ... is that dirty remains long after quick has been forgotten.

Steve McConnell

Triage
Guidelines

Ratcheting
Definition of
Done

Forecasting
Releases

Discerning
Genuine
Unit Testing

Terminology
Definitions

Training
Concerns

Scrum
Master
Selection

Scrum
Diagrams

Official
Scrum
Guide

Virtual
Kanban

J. Carpenter

9   Definition of Done Examples

It is important for an agile team to have a clear, agreed-upon, and explicit quality standard. For Scrum Teams this manifests as a one-page checklist called the Definition of Done.

9.1   DEFINITION OF DONE CRITERIA

A well-crafted definition of Done should achieve each of the following criteria:

- Clearly articulate the quality standard as a checklist of unambiguous statements with simple yes/no answers.

- Stretch the team to achieve the best quality currently achievable.

  » Don't be timid regarding quality. I have seen many engineers teach themselves basic automated unit testing within a day or two. I can say the same about many other technical craftsmanship practices. Good engineers learn most things through self-directed training and practice; please don't underestimate them.

- Avoid currently unachievable criteria.

  » Mixing in aspirational yet currently unachievable criteria prevents the team from seeing the definition of Done as a commitment.

Forging Change

Agile Deployment Models

Agile Design Elements

Mgmt. Behaviors in Scrum

Estimating Business Value

Progressive Refinement at Scale

Sprint Alignment Wall

User Story Ruler

Example Scrum Task Boards

Definition of Done Examples

- Include a version and a change date.

  » The definition of Done should be seen as a constantly evolving quality standard which represents the best the team is currently capable of achieving. The definition of Done represents an agreement between the Scrum Development Team and everyone else, including the Product Owner and stakeholders. If a Scrum Development Team unilaterally relaxes the definition of Done without consulting the Product Owner, it is very likely to damage trust relationships.

- Do not confuse the definition of Done with acceptance criteria on an individual User Story or PBI.

  » Acceptance criteria on individual User Stories are specific to each story, and as such they often vary greatly between stories. In contrast, the definition of Done tends to stay relatively consistent over time. Changes in a definition of Done are usually focused on clarifying existing quality expectations and adding new expectations as a higher quality standard becomes achievable.

Triage
Guidelines

Ratcheting
Definition of
Done

Forecasting
Releases

Discerning
Genuine
Unit Testing

Terminology
Definitions

Training
Concerns

Scrum
Master
Selection

Scrum
Diagrams

Official
Scrum
Guide

Virtual
Kanban

J. Carpenter

- Make the definition of Done rigorous enough to ensure the Sprint Increment is potentially shippable.

  » Early in a Scrum adoption within a legacy water-fall organization there is often a gap between the quality standard required for shipping and what the Scrum Development Team is capable of achieving. I tend to address this issue using a balance of the following two techniques:

    a. Hold the Scrum Development Team to a shippable definition of Done and let the team stumble for a Sprint or two until they rise to the challenge.

    b. Create User Stories that rapidly close the gap between shippable and achievable. Ensure these User Stories are sufficiently prioritized to place them very high on the Product Backlog. See chapter 11, "Ratcheting Definition of Done," for more detail.

Forging Change

Agile Deployment Models

Agile Design Elements

Mgmt. Behaviors in Scrum

Estimating Business Value

Progressive Refinement at Scale

Sprint Alignment Wall

User Story Ruler

Example Scrum Task Boards

Definition of Done Examples

9.2   CONCRETE EXAMPLES

To help make the Definition of Done concept a bit clearer, I have provided a few concrete examples. Other than the first, these are real-world examples with client specifics redacted and generalized. I hope that having examples from more challenging contexts will make it easier for you to see your way through impediments in your own environment.

9.2.1   Context for the First Definition of Done

The first example definition of Done is designed for a typical mature Scrum Team working on a web application with a Java back end and a JavaScript-heavy HTML front end. Most generic Scrum and agile advice you will read is focused on this type of context.

Table 9-1 This simple textbook example assumes a more mature Scrum Team working on a typical Java web application.

First Definition of Done (Ideal Textbook Agile Ecosystem)

Version: 1.23

Date: July 3, 2017

- ❏ Zero bugs at the end of the Sprint within the Dev Team's control to fix.
- ❏ All code changes have been committed to Git master.
- ❏ All relevant builds and tests are passing within the CI environments.
- ❏ Automated unit tests created for any new or modified code, covering both positive and negative test cases regardless of implementation language. This includes unit tests for both JavaScript and Java.
- ❏ Peer review of any new or modified code. Pair programing counts as peer review.
- ❏ All static code analysis errors have been investigated and appropriately resolved.
- ❏ Capable of handling 1000 concurrent transactions per node with no transaction exceeding 20 milliseconds.
- ❏ Variant testing has been considered and accounted for whenever possible.
- ❏ Automated integration tests have been extended as appropriate, and are passing in their entirety.
- ❏ User documentation has been updated as appropriate.
- ❏ Operations documentation has been updated as appropriate.
- ❏ All infrastructure changes are achieved using Infrastructure as Code tooling.
- ❏ Deployment testing successful, including both roll-forward and roll-backward procedures.
- ❏ Log aggregation and performance monitoring integration has been enhanced and validated as appropriate.
- ❏ Database modifications complaint with agreed-upon company standards.
- ❏ Application meets or exceeds corporate security guidelines.
- ❏ Functionality has been deployed to production.
- ❏ All User Story acceptance criteria have been met.
- ❏ Relevant Epic acceptance criteria have been met.

Forging
Change

Agile
Deployment
Models

Agile
Design
Elements

Mgmt.
Behaviors
in Scrum

Estimating
Business
Value

Progressive
Refinement
at Scale

Sprint
Alignment
Wall

User Story
Ruler

Example
Scrum
Task Boards

**Definition of
Done
Examples**

9.2.2   *Context for the Second Definition of Done*

In the second example, a pilot Scrum Team was adding new function-ality to a very large architecturally entangled firmware solution. Since the overall solution was deployed on-premise by tens of thousands of highly risk-adverse customers, absorption cost was exceptionally high. Most customers were only willing to accept an updated release once or twice a year at best. Furthermore, several hundred other engineers were actively modifying the overall product within a waterfall execution model.

You might think success in this situation would be hopeless, yet the team had two key facets of the product working in its favor.

First, from a code-level perspective the functionality being built was almost entirely independent from the rest of the product. There was some overlap with work by other teams, yet the overlap was small enough to be manageable.

Second, the functionality being built could be used in both a stand-alone as well as an integrated mode. Although the integrated solution could only be delivered to customers by hitching a ride on a release of the larger waterfall effort, the stand-alone solution could be shipped any time the Scrum Team was ready. Although the integrated solution was preferable, customers still derived value from incremental enhancements to the stand-alone solution.

This example definition of Done is far from ideal, yet it is probably the best the pilot team was capable of given the impediments it faced. As various impediments were removed, the definition of Done was consis-tently enhanced to embrace a higher quality standard.

Triage
Guidelines

Ratcheting
Definition of
Done

Forecasting
Releases

Discerning
Genuine
Unit Testing

Terminology
Definitions

Training
Concerns

Scrum
Master
Selection

Scrum
Diagrams

Official
Scrum
Guide

Virtual
Kanban

J. Carpenter

Table 9-2 This example comes from a pilot Scrum Team working within a very large scale embedded software environment with hundreds of other teams still executing in a waterfall model for an on-premise deployed product.

Second Definition of Done (Firmware Feature Team in Waterfall Ecosystem)

Version: 2.03

Date: July 5, 2017

Common
- ❑ Zero bugs at the end of the Sprint within the Dev team's control to fix.
- ❑ Any similar functionality between the stand-alone and integrated solution has been factored into common shared libraries.
- ❑ Manually validated against each available hardware platform profile.
- ❑ Release notes updated and published to common location.
- ❑ Configuration information and Release note information saved in source control in a manner which makes it easy for the technical writing team to write documentation for the release.
- ❑ Peer Review completed. (Pair programming counts as peer review.)
- ❑ Demo to Product Owner

Specific to Stand-alone Solution
- ❑ Automated unit test created for both positive and negative cases for any new or modified functionality within custom code. Not required for third-party libraries developed outside the team.
- ❑ Manual Regression executed and passed for all platforms.
- ❑ Passed all solution level test cases for new defects (manual testing today).
- ❑ Final release image build by centralized build systems team is successful.
- ❑ Optimal image size should not increase beyond 100MB.

Specific to Larger Solution
- ❑ Code committed to main branch of overall solution.
- ❑ Automated smoke tests extended to cover any new or modified functionality.
- ❑ Existing commit tests pass.
- ❑ No new static analysis errors introduced.
- ❑ Manual feature test passed.
- ❑ No relevant open defects.
- ❑ No relevant open regression defects.

Forging
Change

Agile
Deployment
Models

Agile
Design
Elements

Mgmt.
Behaviors
in Scrum

Estimating
Business
Value

Progressive
Refinement
at Scale

Sprint
Alignment
Wall

User Story
Ruler

Example
Scrum
Task Boards

Definition of
Done
Examples

9.2.3   Context for the Third Definition of Done

Whenever possible, Scrum Teams should be created as feature rather than component teams. Component teams run a very serious risk of becoming inwardly focused, or at least becoming focused only on the immediate upstream and downstream consuming systems rather than the end customer. In spite of the risk, component teams sometimes still make sense.

The third Definition of Done example is inspired by a group of Scrum Teams responsible for modifying some particularly complex embedded firmware and hardware. The firmware contained a multitude of different subcomponents, each with its own complications which required a great deal of hands-on experience for an engineer to understand in depth. As if that wasn't bad enough, a tremendous amount of the ever-evolving firmware platform code came from an outside vendor that shipped updated code every few weeks with critical updates which could not be ignored. These teams are easily the farthest from a typical middleware software context I have experienced, yet even here the use of Scrum helped facilitate tremendous overall improvements in quality, morale, and forecast accuracy.

With all these challenges and a relatively well-defined component boundary, it seemed to make the most sense to fix engineering practices, cross-training, and overall culture within the component before taking on the additional technical and political challenges creating a feature team would require. As the component Scrum Teams matured, they naturally started pulling in more and more capabilities required to evolve into a feature team.

Triage
Guidelines

Ratcheting
Definition of
Done

Forecasting
Releases

Discerning
Genuine
Unit Testing

Terminology
Definitions

Training
Concerns

Scrum
Master
Selection

Scrum
Diagrams

Official
Scrum
Guide

Virtual
Kanban

J. Carpenter

Table 9-3 This example is inspired by the shared definition of Done used by a group of Scrum Teams working on a very low-level firmware and hardware component team. Generally speaking, organizing as component teams should be avoided in favor of feature teams whenever possible.

Third Definition of Done (Firmware Component Teams in Waterfall Ecosystem)

Version: 1.23

Date: July 3, 2017

- ❏ All User Story acceptance criteria have been met.
- ❏ Relevant Epic acceptance criteria have been met.
- ❏ All activities below performed with latest third-party release bundle.
- ❏ Code peer reviewed in electronic peer review system.
- ❏ No changes outside of pluggable layer, unless agreed by third-party as an exception not handled by pluggable layer.
- ❏ Code checked into official branch.
- ❏ Documentation complete.
- ❏ Features documented via SW Spec Template and checked into document management system.
- ❏ All test cases executed (manual testing allowed).
- ❏ Test plan reviewed.
- ❏ Test management system used to track manual test cases.
- ❏ Zero defects in User Stories (within Dev Team control).
- ❏ Automated release test suite passes.
- ❏ Build is successful on all platforms and automated sanity test passes on all platforms.
- ❏ Legacy defects, or new defects from the waterfall side of the business follow the fire lane process.

Forging
Change

Agile
Deployment
Models

Agile
Design
Elements

Mgmt.
Behaviors
in Scrum

Estimating
Business
Value

Progressive
Refinement
at Scale

Sprint
Alignment
Wall

User Story
Ruler

Example
Scrum
Task Boards

Definition of
Done
Examples

9.3   REFERENCE INFORMATION

A variety of chapter-specific reference information is available on the companion website at http://forgingchange.com/fc_dode. This URL has been encoded in the QR code below for your convenience.

Triage
Guidelines

Ratcheting
Definition of
Done

Forecasting
Releases

Discerning
Genuine
Unit Testing

Terminology
Definitions

Training
Concerns

Scrum
Master
Selection

Scrum
Diagrams

Official
Scrum
Guide

Virtual
Kanban

J. Carpenter

~

There is surely nothing quite so useless as doing with great efficiency what should not be done at all.

Peter F. Drucker

Triage Guidelines

Ratcheting Definition of Done

Forecasting Releases

Discerning Genuine Unit Testing

Terminology Definitions

Training Concerns

Scrum Master Selection

Scrum Diagrams

Official Scrum Guide

Virtual Kanban

J. Carpenter

10 Triage Guidelines

Scrum Teams frequently ask me what they should do about the inevitable "emergencies" which show up and take time away from the Sprint. The challenge is always discerning between real emergencies and perceived emergencies. Any change in focus tends to be extremely disruptive to team productivity, yet real emergencies must still be promptly addressed.

One approach is to route any such demand through the Product Owner, and then allow the Product Owner to determine whether the unplanned work is sufficiently critical to justify deprioritizing ongoing work. This sounds great, and sometimes works. But in practice it tends to be problematic. The Product Owner frequently doesn't know enough to make an informed judgment call without consulting with—and consuming the time of—the Scrum Development Team. Furthermore, the Product Owner is seldom immediately available when the unplanned work shows up.

10.1   FORMALIZED HANDLING OF "EMERGENCIES"

It is usually better to have clearly defined triage guidelines and a process for handling unplanned work. With clearly defined triage guidelines, we empower the people closest to the work to make an effective, immediate decision while still respecting the Product Owner's role in setting business priorities.

Just as important as the ability of a team member to immediately say yes to a critical incoming demand is the ability to say no. People generally want to help each other, and find it hard to say no to a request for help. Formalized triage rules make it much easier for a team member to respectfully say no, while transparently and quickly explaining under what circumstances they can say yes.

The objective isn't to create rigid rules so much as to collaboratively determine and clearly disseminate intent. Team members should have the freedom to use their best judgment, yet be informed by the collective wisdom distilled into the triage guidelines. As with the definition of Done, the triage rules should be reviewed during Sprint Retrospectives and revised as appropriate.

10.1.1 Daily Triage Meetings

It frequently makes sense to create a daily triage meeting or similar mechanism. During this meeting the Product Owner and a few delegates from the Scrum Development Team can quickly sort any demands for unplanned work which didn't meet the criteria for immediate action. This group must be very careful to avoid interrupting the current Sprint work whenever possible.

Daily triage meetings should be kept as short as possible. Five minutes a day just before the Daily Scrum may be sufficient in many cases. With a little creativity, it is sometimes possible to devise a strategy that achieves similar objectives without even needing a meeting.

10.1.2 Tracking Unplanned Work

To help track unplanned work pulled into a Sprint, I recommend creating User Stories and tracking them like any other work. Tracking unplanned work makes it easier to have a more informed discussion with the team and stakeholders about the impact the unplanned work is having on the team during the Sprint Review and Sprint Retrospective. It also makes it possible to track planned versus total Sprint Velocity, and thereby better forecast completion of any upcoming releases.

Triage Guidelines

Ratcheting Definition of Done

Forecasting Releases

Discerning Genuine Unit Testing

Terminology Definitions

Training Concerns

Scrum Master Selection

Scrum Diagrams

Official Scrum Guide

Virtual Kanban

J. Carpenter

Table 10-1 This triage guideline example establishes three basic categories: take immediate action, ask the Product Owner (PO), and put it on the Product Backlog. The Product Owner has clearly communicated intent while empowering the Scrum Development Teams to take immediate action when appropriate. By making the guidelines explicit, the Product Owner also improves the odds of collaboratively refining the guidelines based on the collective wisdom of the teams.

Example Triage Guidelines

Version: 0.33

Date: July 4, 2017

Add to the **Sprint Backlog**, and start work immediately if:
- ❏ I see an email from partner company regarding future platform with subject [Blah-blah] or [Blah-RC]
- ❏ I see an email from build system team, build is broken
- ❏ I see an email indicating the automated long-duration system test failed
- ❏ I see an email from the partner company with Future Platform label notification
- ❏ Platform team believes delaying the work will affect hardware schedule
- ❏ The issue is known to be blocking manufacturing

Add to the **Sprint Backlog**, and start work after approval (PO will likely say yes) if:
- ❏ I get an email from partner company regarding a shipping platform with the subject [Blah-blah] or [Blah-RC] and I then convince myself this is a critical release that needs immediate attention and cannot wait until the next Sprint
- ❏ Severity 1 or Severity 2 bug
- ❏ Issue is raised by manufacturing, but not a blocking item
- ❏ Issue is raised by HW/[Sub-System X]/[Sub-System Y]/[Sub-System Z] teams, but not a blocking item

Add to the **Product Backlog**, we'll prioritize during Product Backlog refinement meetings if:
- ❏ I get an email from partner company with Shipping Platform label notification
- ❏ I get an email with "+Bob" without clear indication of urgency
- ❏ Someone says: "Hi, we need Sam or Sally's help with …"
- ❏ New Severity 3 or lower bug is reported.
- ❏ Asked to work on old Severity 3 or lower bug
- ❏ Platform issue reported, but is not bound to an immediate date

Forging
Change

Agile
Deployment
Models

Agile
Design
Elements

Mgmt.
Behaviors
in Scrum

Estimating
Business
Value

Progressive
Refinement
at Scale

Sprint
Alignment
Wall

User Story
Ruler

Example
Scrum
Task Boards

Definition of
Done
Examples

Many teams visualize the unplanned work using an empty "fire lane" swim lane on their Scrum task boards. Some teams even anticipate a given amount of unplanned work during Sprint Planning and explicitly buffer their velocity to account for it.

10.1.3   Kanban Pull Rules and Triage Guidelines

If you are familiar with Virtual Kanban systems, you will recognize the use of triage guidelines as a simplistic variation on the use of multiple classes of service with complex pull rules, and perhaps multiple input queues.

When truly critical unplanned work begins to dominate the work a team does, greater use of Kanban techniques may be needed. That said, please don't impatiently jump to such a conclusion, as Kanban requires more overhead from a tracking perspective.

When a team can effectively distinguish between perceived and real emergencies, ensures a high-quality definition of Done, and ensures all Scrum Teams have a well-refined Product Backlog, the need for more complex handling of unplanned work usually goes away.

10.2   REFERENCE INFORMATION

A variety of chapter-specific reference information is available on the companion website at http://forgingchange.com/fc_tg. This URL has been encoded in the QR code below for your convenience.

J. Carpenter

**Forging
Change**

Agile
Deployment
Models

Agile
Design
Elements

Mgmt.
Behaviors
in Scrum

Estimating
Business
Value

Progressive
Refinement
at Scale

Sprint
Alignment
Wall

User Story
Ruler

Example
Scrum
Task Boards

Definition of
Done
Examples

*It is not the strongest of the species that
survive, nor the most intelligent, but the
one most responsive to change.*

Charles Darwin

Triage
Guidelines

**Ratcheting
Definition of
Done**

Forecasting
Releases

Discerning
Genuine
Unit Testing

Terminology
Definitions

Training
Concerns

Scrum
Master
Selection

Scrum
Diagrams

Official
Scrum
Guide

Virtual
Kanban

J. Carpenter

11   Ratcheting Definition of Done

Improving the quality standard sometimes requires significant preliminary work. As with any other demand, this preliminary work consumes team capacity. It is therefore frequently a good idea to visualize and manage the effort with a Product Backlog Item (PBI) on the Product Backlog. The relevant PBI is then used to "ratchet" the definition of Done to a higher quality standard.

11.1   CONTEXT

For higher-level software languages, the example Scrum Team was able to immediately incorporate unit testing in their original definition of Done. The substantial learning curve was absorbed by the first User Story. This is often a better approach, although it is usually only practical when the language toolchain being used already has mature unit testing support. For example, a cross-compiled C++ system with complex custom build scripts will likely require far more toolchain changes than a Java system built with Gradle or Maven, which both support JUnit out of the box. The more modern and higher-level the software development language, the better the built-in toolchain support for unit testing is likely to be.

In contrast, the example Scrum Team decided that establishing unit-test infrastructure for their C/C++ component was going to be challenging from a toolchain standpoint. The team decided to use a high-priority User Story on the Product Backlog to deal with raising the quality standard for the code. The relevant User Story is given below.

Forging
Change

Agile
Deployment
Models

Agile
Design
Elements

Mgmt.
Behaviors
in Scrum

Estimating
Business
Value

Progressive
Refinement
at Scale

Sprint
Alignment
Wall

User Story
Ruler

Example
Scrum
Task Boards

Definition of
Done
Examples

Id: US985

Description

As a Development Team member and as a Product Owner, I would like a workable unit test harness for the [Subsystem X] C++ code, so that development efforts will benefit from the fast feedback loops and improved quality unit testing provides.

Acceptance Criteria

- ❑ Every developer on the team who will contribute to the [Subsystem X] code has been ramped up on the new tooling.
- ❑ The definition of Done has been extended to require unit testing for the [Subsystem X] code.
- ❑ Technical solutions for Inversion of Control (likely Dependency Injection), Test Doubles, and Test Harness have been implemented.
- ❑ Sample unit tests have been written to demonstrate use of the framework. These samples should be sufficiently comprehensive to give the Development Team confidence that adding unit testing to the definition of Done will result in an achievable quality standard.
- ❑ The definition-of-Done unit test requirement should be limited to only apply to new or modified code. Any non-trivial changes to a method, function, or class should require unit tests; along with any refactoring required to achieve it. If a modified method calls another unmodified method in an unmodified class, the definition of Done should not mandate unit tests for the unmodified class.

Reference Information

- • http://martinfowler.com/bliki/UnitTest.html
- • http://verraes.net/2015/01/economy-of-tests/

Triage
Guidelines

**Ratcheting
Definition of
Done**

Forecasting
Releases

Discerning
Genuine
Unit Testing

Terminology
Definitions

Training
Concerns

Scrum
Master
Selection

Scrum
Diagrams

Official
Scrum
Guide

Virtual
Kanban

J. Carpenter

11.2   REFERENCE INFORMATION

A variety of chapter-specific reference information is available on the companion website at http://forgingchange.com/fc_rdod. This URL has been encoded in the QR code below for your convenience.

Forging
Change

Agile
Deployment
Models

Agile
Design
Elements

Mgmt.
Behaviors
in Scrum

Estimating
Business
Value

Progressive
Refinement
at Scale

Sprint
Alignment
Wall

User Story
Ruler

Example
Scrum
Task Boards

Definition of
Done
Examples

*We cannot eliminate all variability
without eliminating all value-added.*

Donald Reinertsen,
The Principles of Product Development Flow

Triage
Guidelines

Ratcheting
Definition of
Done

**Forecasting
Releases**

Discerning
Genuine
Unit Testing

Terminology
Definitions

Training
Concerns

Scrum
Master
Selection

Scrum
Diagrams

Official
Scrum
Guide

Virtual
Kanban

J. Carpenter

12   Forecasting Releases

With engineering budgets in the millions, it is usually important to have some idea of what we will get by when. Estimating how much money an engineering team will burn over time is relatively straightforward. Similarly, it is trivial to calculate how many days remain until an important industry event such as a trade show or the beginning of the Christmas shopping season. The tricky part is predicting when a given scope of work will be delivered.

A few agile engineering teams work in extremely dynamic business contexts for which it makes very little sense to spend time planning out work for more than a few Sprints. The need for tight feedback loops for such teams can quickly become an issue of economic survival, not simply optimization of an already profitable venture. These teams will inevitably have frequent deployments to their production environments, and will very rapidly adapt their product direction based on customer feedback.

More commonly, agile engineering teams are working in a business with a well-established business model. Although the business context changes far more rapidly than it may have several decades ago, it is far more stable than that of companies working in an emerging market. In extreme cases the total cost of software engineering efforts is a tiny fraction of the overall development budget. Developing firmware for a new microchip which must be delivered in time for a new chip fabrication plant being constructed or developing the software needed to manage a deepwater oil drilling platform that's under construction are both good examples of extreme delivery date sensitivity.

Even in situations with extreme sensitivity to delivery dates, we must ensure the individual engineer has the personal safety to commit to quality. We must never treat estimates as commitments, because doing so will destroy our ability to accurately forecast completion while concurrently destroying quality and morale.

See "2.4.1 Estimates as Estimates" on page 30 for a detailed explanation of why treating estimates as commitments destroys quality, morale, and forecast accuracy.

Forging
Change

Agile
Deployment
Models

Agile
Design
Elements

Mgmt.
Behaviors
in Scrum

Estimating
Business
Value

Progressive
Refinement
at Scale

Sprint
Alignment
Wall

User Story
Ruler

Example
Scrum
Task Boards

Definition of
Done
Examples

For the sake of clarity I am mostly describing refinement and fore-casting using Scrum terminology even though the concepts are more generally applicable.

The good news is that most macro-level requirements inevitably have a lot of flexibility at the micro-level. By keeping a close eye on progress towards the desired scope, carefully watching historical data, effectively adapting as we learn through experience, and incrementally adjusting scope, an agile team can provide radically better forecast accuracy than a waterfall team. Delivery risk can never be eliminated entirely, but it can be managed effectively—squeezing out as much uncertainty as the nature of the work will allow.

To be clear, I am not saying that fast feedback loops ever stop being important. Engineering practices such as test-driven development, continuous integration, and continuous delivery—and perhaps even continuous deployment and variant testing—are relevant even in cases where forecasting fairly distant releases is important. The focus of this chapter is to help the reader understand release forecasting, not the noble goal of trying to minimize the need to do so.

12.1 MOTIVATIONS FOR PRODUCT BACKLOG REFINEMENT

There are at least two reasons for a Scrum Team to spend time refining the Product Backlog. The first is to achieve commonality of mind with everyone involved; the second is to improve forecast accuracy.

12.1.1 Achieving Commonality of Mind

See chapter 5, "Progres-sive Refinement at Scale" for actionable guidance on achieving common-ality of mind in a large-scale agile environment.

Achieving commonality of mind involves ensuring everyone has a shared understanding of the business intent of each Product Backlog Item (PBI). It also includes helping the engineers anticipate what will be needed from a technical perspective. At times, it becomes evident a PBI should be sliced into multiple PBIs, or that an entirely different approach to solving the business problem should be considered.

When multiple Scrum Teams are working collaboratively, refine-ment is used to anticipate and smooth out any cross-team collaboration issues before they happen. Minimizing cross-team collaboration issues usually requires refining a Product Backlog for at least a couple Sprints in advance.

Triage Guidelines

Ratcheting Definition of Done

Forecasting Releases

Discerning Genuine Unit Testing

Terminology Definitions

Training Concerns

Scrum Master Selection

Scrum Diagrams

Official Scrum Guide

Virtual Kanban

J. Carpenter

Failure to spend the time necessary to achieve commonality of mind usually costs more in terms of lost productivity than the time required to achieve it.

12.1.2   *Improving Forecast Accuracy*

Product Backlog refinement beyond that required for commonality of mind is usually done with the intention of improving the forecast accuracy for an upcoming release. Any such efforts are an implicit choice to reduce production capacity in the hopes of improving forecast accuracy. There is only so much time in the day. Time spent refining a Product Backlog is time not spent doing something else—such as building the actual product.

When refining a Product Backlog beyond a few Sprints, Scrum Teams should heed the typical advice to spend more time on the PBIs closer to the top of the Product Backlog. There is a point of diminishing returns where additional time refining a PBI is unlikely to result in greater forecast accuracy. The farther down the Product Backlog a PBI is, the less time it will take to reach the point of diminishing returns.

When working in an environment with extreme delivery date sensitivity, the cost of improving forecast accuracy past that needed to achieve commonality of mind can make economic sense. The key thing is to recognize that there is a choice to be made, and it should not be made lightly.

Forging
Change

Agile
Deployment
Models

Agile
Design
Elements

Mgmt.
Behaviors
in Scrum

Estimating
Business
Value

Progressive
Refinement
at Scale

Sprint
Alignment
Wall

User Story
Ruler

Example
Scrum
Task Boards

Definition of
Done
Examples

Here is a short list of possible reasons to avoid any refinement efforts beyond those needed to achieve commonality of mind:

- The Scrum Team and the business are constantly learning how to best meet the customer need. The better the ability to quickly validate product direction though frequent incremental delivery, the more true this becomes. The farther down the Product Backlog a PBI is, the greater the chance any refinement detail will be invalidated by new information.

- Reducing production capacity for the sake of forecast accuracy delays product delivery. This delay costs the company revenue it would otherwise have made.

- Delayed delivery slows down the ability to better understand the customer's needs. This increases the risk of wasting time and money building the wrong thing.

Again, I am not saying you should never refine a Product Backlog past the point needed to achieve commonality of mind. I am simply saying that doing so should be a conscious, well-considered decision, not a knee-jerk reaction from a waterfall mindset.

12.2   FORECASTING IN EMPIRICAL PROCESS CONTROL

In complicated systems such as automobile production or microchip fabrication there are well-defined value stream maps consisting of a series of well-understood queue and activity states. This makes it possible to apply sophisticated mathematical techniques to better understand, refine, and forecast the outcomes of such systems. This is not the case for complex systems such as the management of software engineering efforts.

Forecasting when an agile software engineering team will deliver a given scope is a combination of collective team intuition and relatively simple math. Forecast calculations are as simple as dividing the estimated remaining effort by the measured throughput.

Triage Guidelines

Ratcheting Definition of Done

Forecasting Releases

Discerning Genuine Unit Testing

Terminology Definitions

Training Concerns

Scrum Master Selection

Scrum Diagrams

Official Scrum Guide

Virtual Kanban

J. Carpenter

12.2.1 Scrum Forecast Calculations

Most professional Scrum Teams assign a relative effort value measure (e.g., story points) to each PBI towards the top of the Product Backlog. Assuming all PBIs in the Product Backlog within the desired release scope have an assigned effort estimate, a simple sum of these effort estimates divided by the average number of effort units completed per Sprint can be used to calculate how many Sprints likely remain to deliver the desired release scope. The effort units completed per Sprint is commonly referred to as the Sprint Velocity.

Enforcing an assumed throughput is effectively treating estimates as commitments and will destroy forecast accuracy, quality, and morale.

Inevitably, unexpected PBIs will be required for a release. These are a sort of known unknown, frequently termed *dark matter*. Typical guidance is to add around 20 percent to the release scope effort estimate to account for the dark matter. I recommend you also calculate a release forecast using both pessimistic and optimistic Sprint Velocity values.

Forecasting using other agile process frameworks is roughly the same. The calculations inevitably boil down to dividing an estimate of the remaining effort by some form of throughput measurement.

Some teams forgo assigning relative effort estimates to items in their input queue(s). Instead, they calculate throughput on a per-item basis rather than an effort-estimate basis.

Some teams track different types of work independently in addition to tracking them collectively. You will typically see this as a Kanban system with multiple classes of service. Another conceptually equivalent example using different terminology is a Scrum Team which uses a fire lane mechanism governed by established triage rules. The increased tracking detail can improve forecast accuracy, since it becomes easier to understand how much production capacity is likely to be available for building new functionality versus, say, fixing emergency production issues unrelated to the planned release scope. The additional forecast accuracy requires managing additional complexity and process overhead, which may or may not be worth the effort depending on the situation.

12.2.2 Use Measured Throughput Values

The members of most newly formed Scrum Teams without prior experience in a professional Scrum Team are initially overly optimistic about how many PBIs the team can complete per Sprint. After a Sprint or two, the team begins to realize that creating truly shippable Increments meeting a high-quality, potentially shippable definition of Done is far more time-consuming than they expected. At the same time, the team is learning to come together as a team. It will typically take at least four Sprints before things start to stabilize—or, if you prefer, to reach the Norming phase within Tuckman's model of team formation. The stabilization time is somewhat independent of Sprint length, as it is mostly the learning experience of being held accountable for a real-world, high-quality Sprint Increment during the Sprint Review which acts as a forcing function to drive change in perception and behaviors.

Sprint Velocity must be measured; it can not be prescribed. Sprint Velocity is an empirical measure of actual behavior, not desired behavior. Forecasting a release before a somewhat stable Sprint Velocity exists must generally be done using nothing other than collective intuition.

In lean terms Scrum is a pull-based system. The Scrum Development Team pulls PBIs into the Sprint Backlog. The same arguments I am making for using measured throughput when forecasting releases in Scrum also apply to a Kanban-based system.

Triage
Guidelines

Ratcheting
Definition of
Done

**Forecasting
Releases**

Discerning
Genuine
Unit Testing

Terminology
Definitions

Training
Concerns

Scrum
Master
Selection

Scrum
Diagrams

Official
Scrum
Guide

Virtual
Kanban

J. Carpenter

12.2.3   Accept Forecast Variability

I sometimes see management devise a desired Sprint Velocity and then put pressure on the Scrum Development Teams when the actual measured Sprint Velocity is less than the desired value. An equivalent behavior is to put delivery pressure on the Scrum Development Teams when the release forecast is farther out than the business is comfortable with. Both of these behaviors are a vaguely disguised attempt to treat estimates as commitments, and both will destroy forecast accuracy, quality, and morale.

Product managers coming from a waterfall environment sometimes struggle to accept that they must make difficult choices about where to reduce scope. In other words, the business must learn to live within the measured throughput of the engineering system. There are lots of helpful things the business can do to improve the throughput of the system over time, but pretending the physics of the system are different than they are is not one of them.

Forging
Change

Agile
Deployment
Models

Agile
Design
Elements

Mgmt.
Behaviors
in Scrum

Estimating
Business
Value

Progressive
Refinement
at Scale

Sprint
Alignment
Wall

User Story
Ruler

Example
Scrum
Task Boards

Definition of
Done
Examples

12.2.4   Use Uncluttered Velocity Charts

A velocity chart is a simple plot of the number of completed story points per Sprint. With a new Scrum Team (or a Scrum Team working on something unfamiliar), the velocity is likely to be extremely volatile. Over time the velocity will stabilize—and frequently improve, as well.

Electronic application life cycle management (ALM) tools—fancy electronic ticket systems such as Rally—frequently produce velocity charts with a lot of chart junk on them. When communicating outside the team it may be better to craft your own velocity chart. It is important for stakeholders to clearly understand that velocity is a measured value. Noise in a chart increases the chance people will simply ignore the data they probably don't want to accept in the first place.

Triage
Guidelines

Ratcheting
Definition of
Done

**Forecasting
Releases**

Discerning
Genuine
Unit Testing

Terminology
Definitions

Training
Concerns

Scrum
Master
Selection

Scrum
Diagrams

Official
Scrum
Guide

Virtual
Kanban

J. Carpenter

Velocity Chart

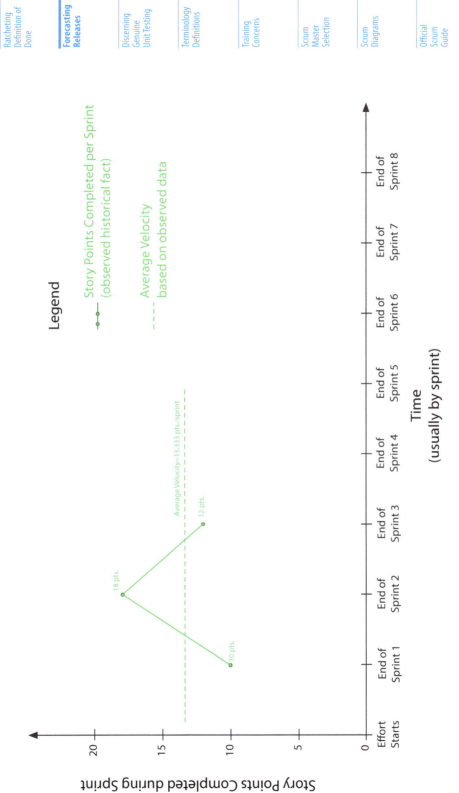

Legend

Story Points Completed per Sprint
(observed historical fact)

Average Velocity
based on observed data

Figure 12-1 A velocity chart is a simple plot of the number of completed story points per Sprint. Charts produced manually and with spreadsheets often have less visual clutter than those produced by ALM tools.

12.2.5 Release Burn-Up Chart

12.2.5.1 Release Burn-Up Chart Benefits

A release burn-up chart is a great way to visualize how forecasts are made in an agile context. It also provides a nice graphical approach for visualizing changes in scope over time.

A release burn-up chart provides insight into how velocity is changing over time, although a velocity chart does so with greater clarity. A velocity chart is equivalent to a chart of the slope of the "Cumulative Story Points Completed" line for the respective release burn-up chart.

Those familiar with Virtual Kanban systems may recognize that a release burn-up chart is a Cumulative Flow Diagram in the degenerate case of a single activity state. The key difference is the overlay of the Planned Release Scope ceiling, which could also be a useful addition to a Cumulative Flow Diagram.

12.2.5.2 Automatic versus Hand-Generated Charts

As with velocity charts, the automatically generated release burn-up charts most electronic ALM tools provide are frequently a bit noisy and devoid of context. A hand-generated release burn-up chart annotated with context is frequently more useful in communicating with stakeholders. (For purposes of this discussion, hand-generated includes release burn-up charts created in a custom spreadsheet and then annotated with contextual detail.)

Release Burn-Up Chart

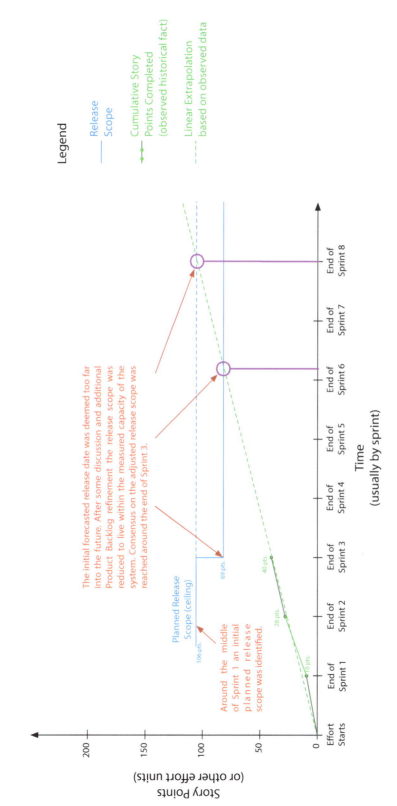

Figure 12-2 A release burn-up chart visualizes historical progress towards a planned release scope along with historical changes to the planned release scope. Charts produced manually and with spreadsheets often have less visual clutter and greater historical context than those produced by ALM tools. A release burn-up chart is similar to a cumulative flow diagram of a process with a single activity state.

Forging
Change

Agile
Deployment
Models

Agile
Design
Elements

Mgmt.
Behaviors
in Scrum

Estimating
Business
Value

Progressive
Refinement
at Scale

Sprint
Alignment
Wall

User Story
Ruler

Example
Scrum
Task Boards

Definition of
Done
Examples

12.3   SCOPING A RELEASE

When initially trying to create a Product Backlog and determine release scope boundaries, techniques such as User Story Mapping and Blitz Planning can be very useful. Jeff Patton's User Story Mapping website content can be found at http://jpattonassociates.com/user-story-mapping/. A description of Blitz Planning can be found on page 68 of *Crystal Clear: A Human-Powered Methodology for Small Teams* by Alistair Cockburn. Cockburn also has a rather sparse web page on Blitz Planning at http://alistair.cockburn.us/Blitz+Planning.

Both Blitz Planning and User Story Mapping are variations on a similar theme. They basically boil down to using various brainstorming techniques to identify and organize all the User Stories in a group setting.

In my experience the knowledge of what is needed to deliver a meaningful business outcome is typically scattered across a large number of people. Efficiently spiraling into a better understanding of what needs to be done generally involves a combination of larger group brainstorming activities and smaller groups working to flesh out details in their area of expertise.

Triage Guidelines

Ratcheting Definition of Done

Forecasting Releases

Discerning Genuine Unit Testing

Terminology Definitions

Training Concerns

Scrum Master Selection

Scrum Diagrams

Official Scrum Guide

Virtual Kanban

J. Carpenter

12.4   ALM TOOL WARNINGS

12.4.1   *Don't Let the Tool Dictate Process*

Be aware that some of the ALM tools do not fully embrace empirical process control. Many of the common enterprise-level ALM tools are sold to customers still firmly entrenched in a waterfall mindset. ALM customers frequently claim to be agile and use lots of agile terminology, yet still mostly behave like a waterfall shop. This reality distorts the ALM marketplace, resulting in money being injected into features and vendors truly agile shops wouldn't need or want.

The legacy of waterfall thinking seems most likely to surface when dealing with ALM support for release planning and forecasting. Releases should always be empirical forecasts based on measured data, not death marches towards fixed scope. To some extent, the larger issue is user intent and understanding of empirical process control, yet it doesn't help when the ALM tool's features and documentation lead a novice user down the wrong path. To the extent practical, make sure you drive the tool rather than get driven by it.

On the upside, most "enterprise" ALM tools I have used are adequate for everything other than forecasting. The older tools such as Rally and Version One are particularly clunky, but even they can be made to work. There are usually more important things to do than to focus too much energy on switching out your ALM tool. I hate to say it, but it is often easier to work around ALM tooling deficiencies with a few spreadsheets than to switch out your ALM tool.

12.4.2   *Choose ALM Tools Carefully*

If you find yourself selecting an ALM tool, I strongly recommend you seek out someone with a great deal of depth in agile methods to help you select a good one. It may also help to have different teams pilot a variety of tools for a few months before picking a winner.

Forging
Change

Agile
Deployment
Models

Agile
Design
Elements

Mgmt.
Behaviors
in Scrum

Estimating
Business
Value

Progressive
Refinement
at Scale

Sprint
Alignment
Wall

User Story
Ruler

Example
Scrum
Task Boards

Definition of
Done
Examples

12.4.3 Understand the Tool's Terminology

You will often have to map the ALM tooling terminology back into your own process. Understanding this mapping is frequently useful as you learn to navigate the tool. As an example, I have included a diagram to help you map Rally's work item queues to those of Scrum.

12.5   REFERENCE INFORMATION

A variety of chapter-specific reference information is available on the companion website at http://forgingchange.com/fc_fr. This URL has been encoded in the QR code below for your convenience.

Triage Guidelines

Ratcheting Definition of Done

Forecasting Releases

Discerning Genuine Unit Testing

Terminology Definitions

Training Concerns

Scrum Master Selection

Scrum Diagrams

Official Scrum Guide

Virtual Kanban

J. Carpenter

Scrum vs. Rally Terminology

Scrum Terminology

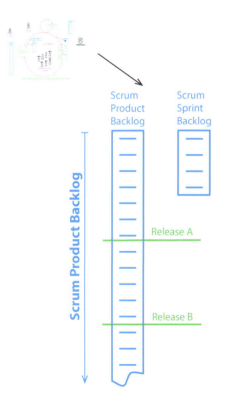

Rally Terminology

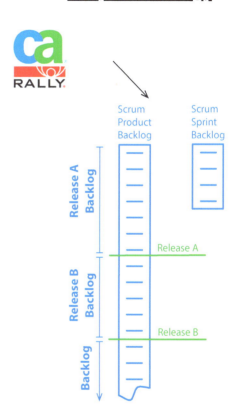

Conceptually the Scrum Product Backlog is a force-ranked queue of every Product Backlog Item not already pulled into the Sprint Backlog. There may be several natural cut lines for releases, yet every PBI is still part of the Product Backlog.

Rally effectively breaks the Scrum Product Backlog into multiple separate sequential queues. Towards the top there is a separate force-ranked queue for each sequential release. Any leftover Rally Work Items not associated with a specific release are considered to be in the Backlog.

Figure 12-3 The terminology of an ALM tool sometimes differs from that of a given agile process framework. As an example, Rally's use of *backlog* is subtly and very confusingly different from Scrum's. This diagram explains how to translate between the two.

Forging Change

Agile Deployment Models

Agile Design Elements

Mgmt. Behaviors in Scrum

Estimating Business Value

Progressive Refinement at Scale

Sprint Alignment Wall

User Story Ruler

Example Scrum Task Boards

Definition of Done Examples

All code is guilty until proven innocent.

Unknown

Triage
Guidelines

Ratcheting
Definition of
Done

Forecasting
Releases

**Discerning
Genuine
Unit Testing**

Terminology
Definitions

Training
Concerns

Scrum
Master
Selection

Scrum
Diagrams

Official
Scrum
Guide

Virtual
Kanban

J. Carpenter

13   Discerning Genuine Unit Testing

Anytime I see teams practicing effective test-driven development (TDD), I inevitably find a variety of other great craftsmanship practices. In contrast, when TDD is missing, overall craftsmanship is inevitably poor in many other areas as well.

Unit testing is a forcing function for well-designed code. The creative processes of crafting elegant production code and elegant unit test code are almost inextricably linked. This isn't about some nebulous, artistic concept of code quality unrelated to business outcomes. Code quality has a major impact on the customer experience, the ability to rapidly add new functionality, and various long-term operational costs.

The combination of a fast compiler and a comprehensive suite of unit tests can detect most problems in seconds. All the other technical feedback loops I am aware of, including automated integration tests, are by themselves generally too slow to keep a large code base from quickly spiraling out of control.

13.1   THE DANGER OF SELF-DECEPTION

In spite of how critical unit testing is and how clearly it has been defined in the agile engineering community, I consistently find tremendous confusion and misuse of the term within companies attempting agile adoptions. Many engineers and most managers in waterfall organizations are loath to admit that what they are calling unit tests are anything but. Managers in these organizations have been unintentionally forcing compromised code quality for years, and have frequently allowed their organization to lose sight of even the most basic craftsmanship practices. It is far easier to redefine unit testing so as to obscure reality than to accept reality and begin to improve. I am not saying these behaviors are necessarily due to willful, conscious deceit. We all have an amazing ability to subconsciously distort our perceptions of uncomfortable facts.

Forging Change

Agile
Deployment
Models

Agile
Design
Elements

Mgmt.
Behaviors
in Scrum

Estimating
Business
Value

Progressive
Refinement
at Scale

Sprint
Alignment
Wall

User Story
Ruler

Example
Scrum
Task Boards

Definition of
Done
Examples

13.2   UNIT TEST DEFINITION

An automated unit test is typically defined as a test focused on validating an individual function or method in isolation. To maintain focus on management behaviors, I will avoid going deeply into technical details; they have been discussed extensively by others. You can find more accurate, nuanced definitions of unit testing in these online articles:

- "UnitTest" by Martin Fowler:
  http://martinfowler.com/bliki/UnitTest.html

- "Economy of Tests" by Mathias Verraes and Konstantin Kudryashov:
  http://verraes.net/2015/01/economy-of-tests

- Wikipedia's "Unit testing" entry:
  https://en.wikipedia.org/wiki/Unit_testing

For a gentler introduction to unit testing, try searching an online bookstore for "Unit Testing." You will likely find a dedicated book tailored to the software development language of your choice.

It is important not to confuse tooling and concepts. Just because an engineer is using this or that unit testing tool doesn't mean proper unit tests are being written. The same tools which facilitate unit testing also help in authoring higher-level tests. That some of these tools have "unit testing" in their name only adds to the confusion.

13.3   DISCERNING TRUTH IN YOUR OWN ORGANIZATION

A decent technical understanding of unit testing is probably the best strategy for discerning the reality in your own organization. Mary and Tom Poppendieck do a great job of detailing the importance of technically competent management in *Leading Lean Software Development*.

> From a lean perspective, the fundamental job of managers is to understand how the work they manage works, and then focus on how to make it better. This is not to say that all leaders must know all the answers; the critical thing is that they *know what questions to ask*. (Italics in original.)

Triage
Guidelines

Ratcheting
Definition of
Done

Forecasting
Releases

**Discerning
Genuine
Unit Testing**

Terminology
Definitions

Training
Concerns

Scrum
Master
Selection

Scrum
Diagrams

Official
Scrum
Guide

Virtual
Kanban

J. Carpenter

Test Pyramid

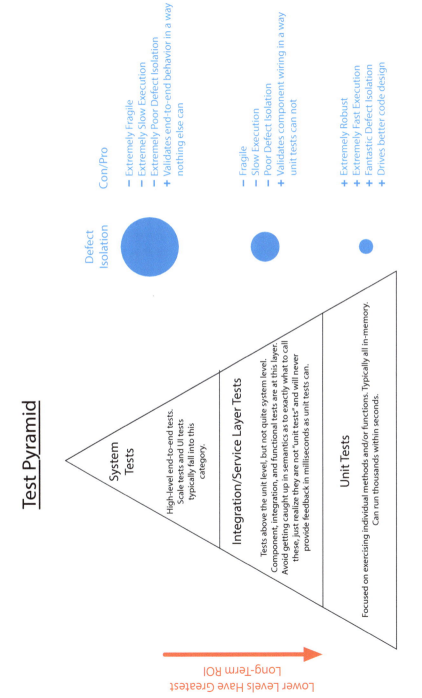

Con/Pro

Defect
Isolation

**System
Tests**

High-level end-to-end tests.
Scale tests and UI tests
typically fall into this
category.

- Extremely Fragile
- Extremely Slow Execution
- Extremely Poor Defect Isolation
+ Validates end-to-end behavior in a way
  nothing else can

Integration/Service Layer Tests

Tests above the unit level, but not quite system level.
Component, integration, and functional tests are at this layer.
Avoid getting caught up in semantics as to exactly what to call
these, just realize they are not "unit tests" and will never
provide feedback in milliseconds as unit tests can.

- Fragile
- Slow Execution
- Poor Defect Isolation
+ Validates component wiring in a way
  unit tests can not

Unit Tests

Focused on exercising individual methods and/or functions. Typically all in-memory.
Can run thousands within seconds.

+ Extremely Robust
+ Extremely Fast Execution
+ Fantastic Defect Isolation
+ Drives better code design

Lower Levels Have Greatest
Long-Term ROI

Figure 13-1 The test pyramid concept is typically used to explain the importance of investing
far more heavily in unit testing than other forms of automated testing. Mike Cohn is generally
credited with introducing the "test pyramid" paradigm in his book *Succeeding with Agile*. The
underlying concepts are far older.

Forging Change

Agile Deployment Models

Agile Design Elements

Mgmt. Behaviors in Scrum

Estimating Business Value

Progressive Refinement at Scale

Sprint Alignment Wall

User Story Ruler

Example Scrum Task Boards

Definition of Done Examples

It is impossible for managers to have detailed technical expertise in every specialization. As long as management knows what questions to ask, it is fine to rely upon the expertise of others. In this case, the negative impact of poor unit testing practices and the level of confusion within most legacy waterfall organizations is too great for a manager to ask the right questions without some level of technical depth. A senior manager must have enough experience with unit testing to see through any smoke screen, or must at least identify a trusted person who can.

Even identifying someone you can trust to properly discern unit test behaviors can be a challenge. I have frequently seen young computer science graduates with a solid understanding of and passion for unit testing working alongside seasoned "architects" in legacy organizations without any significant depth or appreciation of unit testing. These problematic "architects" frequently bear a great deal of responsibility for the current status quo, so be cautious of putting too much faith in their ability to know how to fix things.

Triage
Guidelines

Ratcheting
Definition of
Done

Forecasting
Releases

**Discerning
Genuine
Unit Testing**

Terminology
Definitions

Training
Concerns

Scrum
Master
Selection

Scrum
Diagrams

Official
Scrum
Guide

Virtual
Kanban

J. Carpenter

13.4   SAFETY FOR CRAFTSMANSHIP IS CRITICAL

If engineers are going to ensure elegant production and test code meeting a rigorous quality standard, they must have the personal safety to take the time required to do so. As discussed in chapter 2, it is impossible to lock both time and scope and still produce a well-crafted product. Effort estimates must be treated purely as estimates, while quality should be treated as a commitment.

Until engineers have the personal safety to commit to quality, any focus on improving unit test practices will ultimately fail.

13.5   ESTABLISHING UNIT TEST PRACTICES

Assuming engineers have the personal safety to commit to quality, improved unit test practices are easy to establish.

13.5.1   Raise the Quality Standard and Stand Back in Amazement

Most engineers learn through intentionality and practice. Discussing the topic with the team and requiring unit tests for any new or modified code is often enough. This strategy is most successful when the only barriers to unit testing are cognitive. In many middleware environments the build systems and other tooling already support unit testing.

A middleware team will usually struggle for a few days as they teach themselves basic unit testing while implementing new functionality. With good retrospectives and perhaps some coaching, they will quickly evolve their unit test practices over the course of a few months.

With a little nudging, these same teams will typically start practicing test-driven development soon after they start practicing unit testing. Everything they are likely to read on unit testing best practices tends to encourage test-driven development. Look for opportunities to celebrate the team's craftsmanship improvements, and the team is likely to respond by improving even more.

Take caution to ensure any public accolades are mostly directed at the team as a whole rather than individuals within the team, so as to drive team cohesion and better cultural acceptance of improved craftsmanship.

Forging
Change

Agile
Deployment
Models

Agile
Design
Elements

Mgmt.
Behaviors
in Scrum

Estimating
Business
Value

Progressive
Refinement
at Scale

Sprint
Alignment
Wall

User Story
Ruler

Example
Scrum
Task Boards

Definition of
Done
Examples

There are occasionally a variety of technical barriers which make it much harder to get unit test tooling in place. This is especially true as one moves away from newer software ecosystems in which unit test support has long been the norm. In such cases, putting the necessary tooling in place can easily consume a couple weeks of the team's time. Chapter 11 provides a detailed example User Story inspired by exactly this sort of situation. You can probably leverage the example User Story almost verbatim to help establish unit testing within your own environment.

13.5.2   *Thoughts on Classroom-Based Unit Test Training*

Formal classroom training on unit testing is seldom effective or efficient. Basic unit testing is something most developers can teach themselves relatively easily. Engineers generally prefer to sit down and teach themselves through reading and practice rather than attend formal training. The challenging part of unit testing isn't unit testing itself, but unit testing specific to the application the team is working on. Classroom training seldom provides hands-on experience in the team's current work. Any classroom training is unlikely to be much better than the hundreds of generic references and tutorials available online.

Even though formal classroom training is seldom efficient, pair programing with someone experienced in unit testing is exceptionally useful. This is often accomplished by having a technical coach spend a few days pair programing with the team. I have found it can be almost as effective to have one team with experience coach another. In my personal coaching experience, I generally spend time helping one or two of the more senior engineers and then have them teach everyone else.

This prescription works well when there are only a few teams working on a given code base. But occasionally companies have hundreds of engineers working within a small number of software frameworks. In this narrow situation, formal classroom training can make a lot of sense.

Triage
Guidelines

Ratcheting
Definition of
Done

Forecasting
Releases

**Discerning
Genuine
Unit Testing**

Terminology
Definitions

Training
Concerns

Scrum
Master
Selection

Scrum
Diagrams

Official
Scrum
Guide

Virtual
Kanban

J. Carpenter

The trick to effective training is to ensure that any course involves heavy hands-on student participation developing code in the software frameworks they usually use for work. Creating this kind of training will easily consume several weeks of effort. As an external coach, I find it works best to collaborate with full-time engineers at the client site and let them do the actual course development and delivery. I provide general technical guidance and lend my expertise in instructional design, and ensure the full-time engineers get the accolades.

This advice on formal classroom training is very general. Individual circumstances and specific organizational context might call for a different course of action.

13.6   REFERENCE INFORMATION

A variety of chapter-specific reference information is available on the companion website at http://forgingchange.com/fc_dgut. This URL has been encoded in the QR code below for your convenience.

Forging
Change

Agile
Deployment
Models

Agile
Design
Elements

Mgmt.
Behaviors
in Scrum

Estimating
Business
Value

Progressive
Refinement
at Scale

Sprint
Alignment
Wall

User Story
Ruler

Example
Scrum
Task Boards

Definition of
Done
Examples

*The great enemy of communication ... is
the illusion of it.*

William H. Whyte

Triage
Guidelines

Ratcheting
Definition of
Done

Forecasting
Releases

Discerning
Genuine
Unit Testing

**Terminology
Definitions**

Training
Concerns

Scrum
Master
Selection

Scrum
Diagrams

Official
Scrum
Guide

Virtual
Kanban

J. Carpenter

14   Terminology Definitions

I take a somewhat laissez-faire view of terminology definitions. I am typically content to use whatever terminology is best understood by my conversational partner.

Unfortunately, the terminology around unit testing, continuous integration, continuous delivery, and similar concepts has become increasingly distorted. This distortion is causing confusion around critical concepts. Entirely separate meanings are being conflated in a manner which obscures organizations' understanding of the severity of their gaps in engineering craftsmanship.

Hopefully these definitions will provide conceptual clarity so you can better understand and challenge what you are hearing within your own organization, thereby making it easier to discern the true health of the organization's engineering practices.

Forging
Change

Agile
Deployment
Models

Agile
Design
Elements

Mgmt.
Behaviors
in Scrum

Estimating
Business
Value

Progressive
Refinement
at Scale

Sprint
Alignment
Wall

User Story
Ruler

Example
Scrum
Task Boards

Definition of
Done
Examples

14.1   DEFINITIONS

Unit Test

An automated test focused on validating an individual function, method, or class. As a unit test typically avoids both disk and network I/O, you can generally run thousands in a few seconds or less.

Unit Test Framework

Automated test harnesses such as JUnit that are intended to help software engineers easily organize and execute automated tests. Not all tests written within a unit test framework are necessarily unit tests. There is nothing wrong with using a unit test framework when building higher-level automated tests. Just don't muddy the concepts of a unit test and a unit test framework.

Continuous Integration as a Behavior

Continuous integration as a behavior requires engineers to be **constantly merging to a single source of truth**, and ensuring this single source of truth always passes base-level sanity tests.

As engineers, we communicate via our code as much as we do through any other mechanism. If we are not merging our changes to a consolidated, validated single source of truth on a regular basis, we are not communicating.

Practicing continuous integration as a behavior is about discipline, not tooling. James Shore's article *"Continuous Integration on a Dollar a Day"* (http://www.jamesshore.com/Blog/Continuous-Integration-on-a-Dollar-a-Day.html) describes a very simple approach for implementing continuous integration as a behavior, even without sophisticated tooling.

The practice and discipline of Continuous Integration as a Behavior is far more important to efficient delivery of customer value than Continuous Integration as a Tool.

Triage
Guidelines

Ratcheting
Definition of
Done

Forecasting
Releases

Discerning
Genuine
Unit Testing

**Terminology
Definitions**

Training
Concerns

Scrum
Master
Selection

Scrum
Diagrams

Official
Scrum
Guide

Virtual
Kanban

J. Carpenter

Continuous Integration as a Tool

Continuous Integration as a Tool facilitates continuous integration (CI) by watching for source control changes, executing a command-line build and test harnesses against the changes, and reporting on results. Failures typically trigger alerts to the engineers. Common examples of CI tooling include Jenkins, Team City, and Bamboo.

CI tools are wonderful, but they are no replacement for Continuous Integration as a Behavior. Humans are fantastically creative, and able to work around any attempt to use a tool as a means to force one behavior or another.

Just because a team is using a CI tool such as Jenkins or TeamCity does not mean they are practicing Continuous Integration as a Behavior. Many teams have an automated CI tool in place yet pay no attention to failures. Similarly, many teams have an automated CI tool monitoring a large number of source control branches, yet fail to integrate the branches into a single validated mainline multiple times per day.

The use of Continuous Integration as a Tool is not nearly as important to efficient delivery of customer value as the practice and discipline of Continuous Integration as a Behavior.

Continuous Delivery as a Behavior

Continuous Delivery as a Behavior is the practice of continually creating fully validated potentially shippable release candidates. From an engineering perspective, the focus is on continually driving out all technical risk.

Just because the release candidate is fully validated from a technical perspective does not mean it will actually ship to production. Whether the additional functionality available differs enough to justify the over-head of shipping (i.e., absorption and other transaction costs) is a business decision.

Forging
Change

Agile
Deployment
Models

Agile
Design
Elements

Mgmt.
Behaviors
in Scrum

Estimating
Business
Value

Progressive
Refinement
at Scale

Sprint
Alignment
Wall

User Story
Ruler

Example
Scrum
Task Boards

Definition of
Done
Examples

Continuous Delivery as a Behavior should not be confused with Continuous Deployment as a Behavior. Continuous Deployment as a Behavior implies the release candidate is continually shipped, whereas Continuous Delivery as a Behavior only implies the release candidate *could* be shipped.

Continuous Delivery as a Behavior naturally implies Continuous Integration as a Behavior, but the reverse is not necessarily true.

If you read the definition of the Increment in the Scrum Guide, you will realize it effectively requires the Scrum Development Team to practice Continuous Delivery as a Behavior.

Continuous Deployment as a Behavior

Continuous Deployment as a Behavior is the practice of continually shipping fully validated potentially shippable release candidates.

Continuous Deployment as a Behavior assumes that the overhead of shipping has been reduced to the point that continually shipping the release candidates is a practical choice.

Continuous Deployment as a Behavior naturally implies Continuous Delivery as a Behavior, but the reverse is not necessarily true.

Continuous Delivery as a Tool

Continuous Delivery as a Tool is the natural extension of Continuous Integration as a Tool, with the addition of fully automated deployment and validation of the release candidates through a series of test environments.

Continuous Delivery as a Tool implies Continuous Integration as a Tool, but the reverse is not necessarily true.

In many cases, Continuous Delivery as a Tool will incorporate Infrastructure as Code concepts to facilitate environment management.

Triage Guidelines

Ratcheting Definition of Done

Forecasting Releases

Discerning Genuine Unit Testing

Terminology Definitions

Training Concerns

Scrum Master Selection

Scrum Diagrams

Official Scrum Guide

Virtual Kanban

J. Carpenter

Continuous Deployment as a Tool

Continuous Deployment as a Tool is practically identical to Continuous Delivery as a Tool. To perform Continuous Deployment as a Behavior within a CD tool, one simply configures the CD tool not to bother asking a human for permission before deploying to production.

Infrastructure as Code

Infrastructure as Code tools allow the expression of a desired infrastructure state as code maintained in a source control system. Consequently, any change to the infrastructure can flow down an automated Continuous Delivery as a Tool build pipeline, just as any changes to normal system code do. This contrasts with the traditional use of manually executed "run-books" maintained in MS Word documents, wiki pages, production change tickets, etc.

Although Infrastructure as Code can be implemented purely as an engineering design pattern, there are many tools which attempt to simplify the task. These tools typically separate declaration of desired intent from behavior needed to achieve that intent. Often, they provide pre-built modules which attempt to implement the desired intent across a large variety of operating systems. Unfortunately, these tools tend to be focused on individual nodes, so one still tends to spend time writing orchestration code as well.

Examples of Infrastructure as Code tools include Puppet, Chef, Ansible, and Salt-Stack.

In this case, "code" means infrastructure-focused text files checked into source control. For example, within a Puppet configuration one might use the "file" resource to ensure a specific file exists with specific file permissions and content.

```
#Example Puppet configuration snippet

file { '/etc/be_happy':
  owner => 'root',
  group => 'root',
  mode => '0644',
  content => "hello, world!\n",
}
```

If this is all a bit too abstract, I recommend you read *Continuous Delivery* by Jez Humble and David Farley. Although the tooling in *Continuous Delivery* is a bit dated, the concepts remain very relevant.

14.2   REFERENCE INFORMATION

A variety of chapter-specific reference information is available on the companion website at http://forgingchange.com/fc_td. This URL has been encoded in the QR code below for your convenience.

I hear and I forget. I see and I remember.
I do and I understand.

Confucius

Triage
Guidelines

Ratcheting
Definition of
Done

Forecasting
Releases

Discerning
Genuine
Unit Testing

Terminology
Definitions

**Training
Concerns**

Scrum
Master
Selection

Scrum
Diagrams

Official
Scrum
Guide

Virtual
Kanban

J. Carpenter

15   Training Concerns

Ensuring that everyone understands your chosen agile execution model improves communication with the organization and helps set role expectations. The challenge is how to achieve this.

Some level of formal classroom training at multiple levels of a large organization makes sense. But classroom training has an exceptionally high opportunity cost in terms of participant time, and should be used very judiciously. It's important to ensure the instructional design and delivery achieves long-term retention and provides actionable guidance.

15.1   CLASSROOM TRAINING IS OFTEN THE WRONG CHOICE

In my experience most professionals learn through self-directed study and practice. Organizations frequently try to achieve through formal classroom training that which is best learned by experience and practice. Before making people spend a few days in a formal classroom setting, carefully consider whether there are more efficient ways achieve the learning objectives.

You may also want to analyze why the selected participants have not already learned the relevant topic. If structural obstacles have prevented certain desired behaviors, removing the structural obstacles may be all you need to do.

Forging
Change

Agile
Deployment
Models

Agile
Design
Elements

Mgmt.
Behaviors
in Scrum

Estimating
Business
Value

Progressive
Refinement
at Scale

Sprint
Alignment
Wall

User Story
Ruler

Example
Scrum
Task Boards

Definition of
Done
Examples

15.2   CLASSROOM TRAINING DESIGN

The true measure of effective training is how much of the training content people retain over time and whether they are able to put it into practice. It is better to deliver less content in a way participants will retain than more content without the exercises needed to achieve retention.

Based on my experience designing, delivering, and attending formal classroom training, effective learning outcomes are best achieved when a course is designed with a great deal of deep, meaningful, hands-on participation. Everything I have read from experts in modern instructional design reinforces this view. In the words of Dr. Sivasailam "Thiagi" Thiagarajan, an instructor should "be a sage by the side" as opposed to a sage on stage.

If you have multiday training in which students mostly listen to an instructor deliver a lecture while going through a lot of slides you have a problem. Lecture and slides can both make sense in moderation, but if they dominate the training it is very unlikely participants will retain much.

Like many other things, effective instructional design and delivery is a learned skill. Two of my favorite experts on applied instructional design are Thiagi and Bob Pike. Sharon L. Bowman is also quite popular.

Triage
Guidelines

Ratcheting
Definition of
Done

Forecasting
Releases

Discerning
Genuine
Unit Testing

Terminology
Definitions

Training
Concerns

Scrum
Master
Selection

Scrum
Diagrams

Official
Scrum
Guide

Virtual
Kanban

J. Carpenter

15.2.1 Customized Training

The closer training is to real-world application, the better the learning outcomes tend to be. Ideally, training is customized to the specific context the participants are working in. As an example, it is better if large-scale Product Backlog refinement training uses the application life cycle management tool and project structure that participants will use outside of training. Similarly, ideal unit test training will involve authoring tests and code in the code base and toolchain the participants normally use.

A more nuanced discussion of unit test training can be found in "13.5.2 Thoughts on Classroom-Based Unit Test Training" on page 142.

Unfortunately, creating context-specific training is extremely time-consuming. Well-designed training frequently takes a few weeks to build, and it usually takes a few rounds of delivering the training and rolling in feedback to polish a course. The lightweight instructional design approaches advocated by Thiagi help minimize course development time, yet even his techniques are not a magic bullet.

In practice, one must balance the benefits of customized training with the opportunity costs of spending coaching resources elsewhere. The broader the audience to which the training is relevant, the more likely it will make sense to create custom training. A few relevant factors include the cost of delaying organizational improvement until more ideal training can be created, course development labor costs, the number of participants who must endure less optimal training, and the costs of delaying validation of the course design.

15.2.2 Ad Hoc Training

In a small enough group, ad hoc training can make a lot of sense. A good instructor who knows the subject can do pretty well with nothing other than a whiteboard and a bunch of sticky notes. The learning outcomes are unlikely to be as good as more carefully designed training, yet they may be acceptable.

Forging Change

Agile Deployment Models

Agile Design Elements

Mgmt. Behaviors in Scrum

Estimating Business Value

Progressive Refinement at Scale

Sprint Alignment Wall

User Story Ruler

Example Scrum Task Boards

Definition of Done Examples

When there is little development time available and appropriate written content exists, some of Thiagi's textra frame games often work very well. If the written content doesn't exist, someone still has to write it. As you might have guessed, a secondary design goal for some of the reference content in this book is to serve as input text for use in classroom training.

15.3   THE TIMING OF TRAINING

When newly acquired knowledge is not quickly applied, it is often forgotten. Well-designed immersive training greatly improves long-term retention, but it isn't magic. The sooner participants are applying what they have learned the better.

When standing up agile teams, it is best to use classroom training to establish baseline expectations and stand up the teams within a week or so of the training. If it has been more than a few months since a group has been trained, it is probably best to at least have a refresher course before standing them up as an agile team.

I frequently see companies send a large number of people through agile training only to return them to the same environment they came from. At best, participants forget most of what they have learned within a few months. At worst, participants become more aware of and depressed by dysfunctions in their own environment and feel disheartened by lack of actionable support in their management chain to improve things.

One sensible exception is helping people who will continue to execute in a traditional fashion while interacting with agile teams. Providing these stakeholders a basic understanding of what to expect from the agile teams will help everyone communicate better. Even this sort of training is best delivered as close in time to the actual interactions as possible. As long as the agile teams are executing well, the stakeholders may learn all they need simply by interacting with them.

Triage
Guidelines

Ratcheting
Definition of
Done

Forecasting
Releases

Discerning
Genuine
Unit Testing

Terminology
Definitions

**Training
Concerns**

Scrum
Master
Selection

Scrum
Diagrams

Official
Scrum
Guide

Virtual
Kanban

J. Carpenter

15.4   MANAGEMENT MUST LEARN AS WELL

As discussed in chapter 1, management bears responsibility for establishing a structure in which the desired culture can flourish. Managers therefore have an obligation to those they have the privilege to lead, as well as to their employer, to gain an even deeper understanding of agile structural elements than anyone else.

If a Scrum Development Team is entirely accountable for delivery of an Increment meeting the definition of Done, then what is there for an engineering manager to do other than help to improve the system? How can a manager do this without understanding the nature of the work or the process framework being used? As I said in chapter 1, most of the problems with engineering teams are a direct result of management decisions and behaviors.

15.5   AGILE COACH RELATIONSHIPS

One of the challenges as an external consultant is developing rapport with those I am trying help. Delivering classroom training provides an excellent opportunity to build a better relationship with the participants and to be seen as an expert on the topic. When I'm helping to stand them up as an agile team a few days later, these relationships make it much easier to get the job done.

Clients understandably want to make efficient use of expensive consultants. Sometimes, I see companies try to centralize training in an effort to improve efficiencies. Even if this approach optimizes training efficiency, it generally reduces the overall effectiveness of coaching efforts. I also find centralized training usually devolves, as large companies inevitably end up trying to save money on training by having less qualified instructors deliver it.

It is usually best to have the person coaching the team deliver any formal classroom training. If the agile coach is less experienced, consider having them co-train with a more seasoned coach.

Forging
Change

Agile
Deployment
Models

Agile
Design
Elements

Mgmt.
Behaviors
in Scrum

Estimating
Business
Value

Progressive
Refinement
at Scale

Sprint
Alignment
Wall

User Story
Ruler

Example
Scrum
Task Boards

Definition of
Done
Examples

15.6   SPONSORSHIP SUPPORT

An external consultant typically has no significant positional authority within the client company. Training participants know that what they are being taught only matters to the extent their own management is willing to enact it. So it's a good idea for the executive sponsor or similar senior manager to kick off any formal classroom training, taking a few moments to emphasize the importance of the course content and introduce the instructor.

15.7   CERTIFIED TRAINING GAME

Creating meaningful organizational change is very hard. At any one time, there are not many executives seeking the sort of change an agile adoption entails. In contrast, the illusion of change is very seductive and requires much less struggle to achieve.

The market forces in the agile community have driven a large demand for agile certifications and the associated certified training. Worse yet, some of the most well-known certifications have exceptionally low barriers to entry. Ironically, the more comprehensive and extensive a certification becomes, the lower the market demand for those with the certification seems to be.

I have worked with countless people claiming to be Scrum Masters who have a corresponding Scrum Master certification. Very seldom do these individuals have more than a surface-level understanding of agile fundamentals. Even people with supposedly more advanced certifications frequently are not much better. The most common anti-pattern I see is people who perceive agile process frameworks as a means of achieving greater transparency in order to better micromanage engineers.

I am not saying all certifications are bad; I am just pointing out how murky the market for certified agile training has become.

Triage Guidelines

Ratcheting Definition of Done

Forecasting Releases

Discerning Genuine Unit Testing

Terminology Definitions

Training Concerns

Scrum Master Selection

Scrum Diagrams

Official Scrum Guide

Virtual Kanban

J. Carpenter

15.8   FOLLOW-ON TRAINING

Many of the nuanced aspects covered in agile training will be lost on people without any real-world experience working within an agile team. It can therefore be useful to provide additional follow-on training a few months after standing up an agile team. A set of half-day classroom trainings spread over several weeks or a study group format can work well for this type of training.

15.9   REFERENCE INFORMATION

A variety of chapter-specific reference information is available on the companion website at http://forgingchange.com/fc_tc. This URL has been encoded in the QR code below for your convenience.

Forging Change

Agile Deployment Models

Agile Design Elements

Mgmt. Behaviors in Scrum

Estimating Business Value

Progressive Refinement at Scale

Sprint Alignment Wall

User Story Ruler

Example Scrum Task Boards

Definition of Done Examples

When a flower doesn't bloom you fix the environment in which it grows, not the flower.

Alexander den Heijer

Triage
Guidelines

Ratcheting
Definition of
Done

Forecasting
Releases

Discerning
Genuine
Unit Testing

Terminology
Definitions

Training
Concerns

**Scrum
Master
Selection**

Scrum
Diagrams

Official
Scrum
Guide

Virtual
Kanban

J. Carpenter

16   Scrum Master Selection

Traditional waterfall shops typically have separate and deep management hierarchies for quality assurance and development staff. The point in the organizational chart where these management hierarchies converge is frequently many layers up from the individual contributors. The same can be said for the management hierarchies of project managers.

A Scrum Development Team is designed as a flat structure in which the team is collectively held accountable for delivering an Increment meeting the definition of Done. Creating a Scrum Development Team with the cross-functional skill set capable of delivering a shippable Increment typically requires assembling people who traditionally were working in these separate organizational silos.

So long as all the various management chains are well aligned and the Scrum Development Team members are given consistent direction aligned with Scrum, it doesn't really matter who they report to. There is even a case to be made for aligning individual contributors with managers of similar technical backgrounds. Unfortunately, the various management chains are seldom well aligned. This causes a variety of problems which result in Scrum Development Team members being stretched between the expectations of their direct manager—who controls their performance review —and the Scrum Master. Similarly, the direct managers frequently have trouble respecting the Scrum Product Owner's authority to prioritize work consuming the production capacity of the Scrum Development Team.

The best long-term solution is to revamp the organizational chart and the various management role definitions to avoid these conflicts. Unfortunately, there is seldom sufficient political capital to effect such change early in an agile adoption. Senior management is often reasonably fearful of making too many big changes at once.

Forging
Change

Agile
Deployment
Models

Agile
Design
Elements

Mgmt.
Behaviors
in Scrum

Estimating
Business
Value

Progressive
Refinement
at Scale

Sprint
Alignment
Wall

User Story
Ruler

Example
Scrum
Task Boards

Definition of
Done
Examples

16.1   SCRUM STUDIO LEADER SELECTION

Any serious Scrum adoption usually starts with some flavor of the Scrum Studio Change Model. This is designed to test out the new operating model in a small pilot before making broader organizational changes. Establishing these Scrum pilot teams frequently requires working within a traditional organizational structure at first.

Deciding what will work best in a particular setting usually comes down to weighing the strengths and weakness of individuals available for the roles needed. Even if the pilot Scrum Teams are set up in a completely separate portion of the organizational chart, the people still have to come from somewhere. As such, their career history probably has influenced their strengths and weaknesses.

16.2   LOOK FOR SKILLS, NOT TITLES

In trying to provide overall leadership selection guidance I will speak in broad, general strokes. We are all somewhat a product of our environment. Individuals who have spent their entire career in quality assurance within a waterfall organization will tend to see the world differently from others who have spent their career working in software engineering, project management, or system administration.

Perhaps it is best to think of my descriptions in terms of skill-set groups that just happen to align with the traditional silos in a waterfall organization. An organization is built of people, not automatons, so it is important to remember that each candidate should be evaluated as an individual whose skills may vary from the norm.

For the purposes of this chapter, I am intentionally risking overgeneralization in the hope of providing actionable guidance. I hope you will use a bit of common sense in applying this advice, paying more attention to the spirit of the discussion than the exact details.

Triage Guidelines

Ratcheting Definition of Done

Forecasting Releases

Discerning Genuine Unit Testing

Terminology Definitions

Training Concerns

Scrum Master Selection

Scrum Diagrams

Official Scrum Guide

Virtual Kanban

J. Carpenter

Scrum Leadership Role Choices If Avoiding Major Restructuring

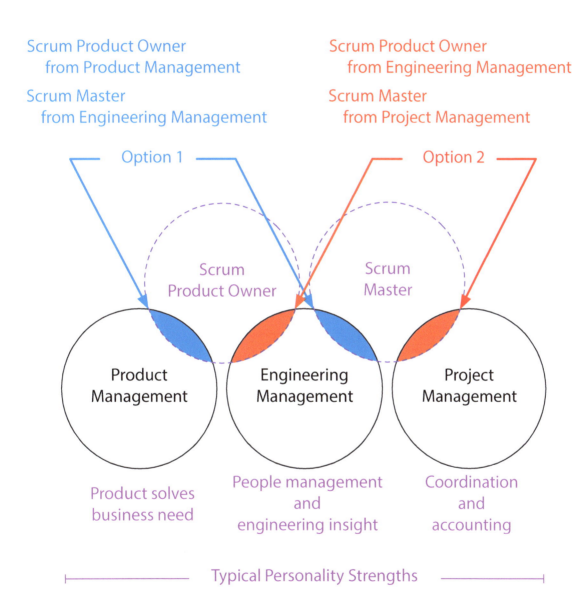

Figure 16-1 The professional strengths and weaknesses of leadership candidates in a traditional organization are heavily influenced by their past experience within the organization. Any leadership choice made when establishing a pilot Scrum Team is inevitably a compromise. The figure on the next page elaborates on the two typical choices shown here.

Comparative Advantages of Scrum Leadership Role Choices
(while avoiding major restructuring)

Agile Deployment Models

Agile Design Elements

Mgmt. Behaviors in Scrum

Estimating Business Value

Progressive Refinement at Scale

Sprint Alignment Wall

User Story Ruler

Example Scrum Task Boards

Definition of Done Examples

Option 1

Scrum Product Owner from Product Management

Scrum Master from Engineering Management

Option 2

Scrum Product Owner from Engineering Management

Scrum Master from Project Management

Option 2:
- SM safe transparency 2A
- PO staff available 2B
- SM enforces structure 2C

Option 1:
- managers serve team 1A
- SM has eng. background 1B
- PO responsible for P&L 1C
- PO is not a dev. mgr. 1D
- biz controls production 1E

An advantage for one option is a disadvantage for the other.

164

Triage
Guidelines

Ratcheting
Definition of
Done

Forecasting
Releases

Discerning
Genuine
Unit Testing

Terminology
Definitions

Training
Concerns

**Scrum
Master
Selection**

Scrum
Diagrams

Official
Scrum
Guide

Virtual
Kanban

OPTION 1 LEGEND

Managers serve team[1A]

The primary role of engineering management is to serve those they lead. The Scrum Development Team, not management, creates the work product the end customer cares about.

SM has eng. background[1B]

The official Scrum Guide's description of the ideal Scrum Master seems closest to an engineering manager practicing servant-leadership, with deep knowledge of agile process and software craftsmanship. Project managers seldom have enough technical depth to identify the sort of challenges an engineering manager can, nor the same type of experience and relationships needed to solve them.

PO responsible for P&L[1C]

Product Owner is effectively responsible for profit and loss. It is a business role.

PO is not a dev. mgr.[1D]

There is a natural conflict of interest in having members of a Scrum Development Team report to the Product Owner. There is a strong temptation for a Product Owner to emphasize the pressing needs of product delivery over minimizing the less visible long-term penalties of poor software craftsmanship.

Biz controls production[1E]

Traditionally, engineering managers are empowered to manage the production capacity of the engineers. This can be culturally challenging for engineering managers to overcome, even when they are self-aware enough to see the issue. Having a Product Owner outside of engineering helps to correct this.

OPTION 2 LEGEND

SM safe transparency[2A]

The Project Management Organization in a waterfall organization already has an orthogonal reporting structure within waterfall, which was put in place to help ensure transparency in a waterfall environment. Leveraging project managers as Scrum Masters makes it possible to obtain transparency without significant change to the organizational structure. Although the better long-term solution is to make more radical organizational changes, the short-term productivity impact may be too expensive.

PO staff available[2B]

Product Management in a waterfall organization seldom has enough staff to effectively provide Product Ownership support for a large number of Scrum Development Teams.

SM enforces structure[2C]

Project managers are usually exceptionally good at ensuring meetings occur and stay focused.

Figure 16-2 The diagram above contrasts the typical advantages of the two leadership selection options identified in figure 16-1. The details of each advantage are then further detailed within the legend. These characteristics highlight the contrast between the choices. The two options shown are a gross oversimplification of diverse real-world options, yet will hopefully bring more nuance to the surface and help produce more carefully considered leadership choices.

Forging Change

Agile Deployment Models

Agile Design Elements

Mgmt. Behaviors in Scrum

Estimating Business Value

Progressive Refinement at Scale

Sprint Alignment Wall

User Story Ruler

Example Scrum Task Boards

Definition of Done Examples

16.3   AVOID WATERING DOWN THE SCRUM MASTER ROLE

Engineers typically deal with tremendous complexity in their day-to-day work. In comparison to other topics they must master, learning to run Scrum well simply isn't that hard. In a well-run Scrum Team, the cross-functional Scrum Development Team takes full accountability for delivery. Solving problems the Scrum Development Team is able to solve for itself disempowers the Scrum Development Team.

I commonly see individuals with the skill set of a junior program manager assigned as Scrum Master to one or two Scrum Teams. These individuals are usually unable to remove obstacles the Scrum Development Team can't remove on its own. They lack both the positional authority to address organizational obstacles and the engineering expertise to understand and remove technically detailed obstacles. Since these individuals often lack the nuanced interpersonal skills managers have spent so much time developing, they are seldom effective at resolving difficult personality conflicts within the Scrum Team.

A similar anti-pattern is a Scrum Development Team member who acts as a Scrum Master. In this case the Scrum Master is still no more able to remove obstacles than the Scrum Development Team itself.

Please understand, I have great respect for people with an effective project management skill set. I know project managers who are fantastic at working across the organization to pull things together. Effective project managers are also fantastic to have as members of any Scrum Development Team which is constantly interacting and integrating with outside companies.

Some of the most impressive project managers I have ever worked with are at a client producing custom enterprise server hardware. These individuals keep an endless stream of complex prototype hardware flowing into and out of the software engineering teams. Without these project managers working several months ahead of the software teams the

Triage
Guidelines

Ratcheting
Definition of
Done

Forecasting
Releases

Discerning
Genuine
Unit Testing

Terminology
Definitions

Training
Concerns

**Scrum
Master
Selection**

Scrum
Diagrams

Official
Scrum
Guide

Virtual
Kanban

J. Carpenter

entire engineering group will quickly grind to a halt. Their manager has a professional background in production process control, and knows more about detailed statistical analysis of mass production plants than I will likely ever need to learn.

In summary, don't expect to assign a role requiring the skills and authority of a seasoned manager to a junior employee and be successful. Professional, seasoned engineers deserve a Scrum Master they can respect. Please choose wisely.

16.4   LEADERSHIP SELECTION GUIDANCE

With the above context in mind, please stop and reread the Scrum Master role definition in the official Scrum Guide. An engineering manager who long ago learned not to micromanage their reports and has always focused on serving others is probably a better Scrum Master candidate than your typical project manager. This individual will likely lack expertise in empirical process control, but that is relatively straightforward to teach. There may also be an initial struggle to accept that they are no longer accountable for delivery, but a combination of cultural and structural leadership in the executive management layer can quickly fix this as well.

My definitions of cultural leadership and structural leadership can be found on page 4 and page 5.

Teaching an engineering manager to be an effective Scrum Master is straightforward and relatively simple compared to other things they have already mastered. Even so, a deep understanding still requires more than a couple days in a classroom setting. The sort of depth I like to see usually requires an extended study program consuming a few hours per week over the course of six months or more, combined with hands-on immersion in an agile environment. An engineering manager has typically spent years as a hands-on software engineer and is therefore better able than most to connect the dots between theory and practice. I recognize this is a biased view. I have seen people from other backgrounds with impressive depth in empirical process control; it is simply less common.

Forging Change

Agile Deployment Models

Agile Design Elements

Mgmt. Behaviors in Scrum

Estimating Business Value

Progressive Refinement at Scale

Sprint Alignment Wall

User Story Ruler

Example Scrum Task Boards

Definition of Done Examples

In contrast, you may have an engineering manager who has always micromanaged everyone and constantly sought out the spotlight. Please find someone else to act as the Scrum Master. My knee-jerk reaction is to ensure such a person is kept far away from any Scrum studio efforts. These attitudes and behaviors can always be dealt with later, when the organization is ready for the Executive Pull–Based Change Model. In my experience, even observing successful teams from a distance can change and soften entrenched behaviors. For the Scrum Master role, I would rather have a senior effective project manager who has embraced servant-leadership than the problematic engineering manager described above.

Companies often find they don't have enough people qualified to act as a Product Owner. Finding a Product Owner can be especially problematic with a pilot Scrum Team that is more of a component team than a feature team. I strongly recommend initially focusing on feature teams for a Scrum studio, although this isn't always possible. In such cases, you are likely to find the engineering manager is the only person with sufficient technical knowledge to act as the Product Owner. In this situation a seasoned project manager may be the best Scrum Master choice available. Having the same person play the Scrum Master and Product Owner role is possible but risky and best avoided, especially in a pilot effort.

For sake of clarity, please remember the Scrum Master is not necessarily a functional manager with direct reports. The Scrum Master role is a management role responsible for process. A person playing the Scrum Master role may or may not also be playing a functional manager role. The Scrum Master role is orthogonal to a functional manager role. In my experience a candidate with the depth and breadth of experience necessary to be successful as a Scrum Master will also have the depth and breadth of experience to be successful as a functional manager.

Triage Guidelines

Ratcheting Definition of Done

Forecasting Releases

Discerning Genuine Unit Testing

Terminology Definitions

Training Concerns

Scrum Master Selection

Scrum Diagrams

Official Scrum Guide

Virtual Kanban

J. Carpenter

16.5   FACILITATING NUANCED LEADERSHIP DISCUSSION

It is extremely difficult to have a nuanced discussion regarding leadership selection without first establishing a common baseline of understanding. It also helps to establish a common vocabulary for efficient communication. I hope this chapter helps in that regard.

Before discussing leadership selection, consider having everyone involved in the decision read this chapter and the role definitions in the official Scrum Guide first. You might also search out other complementary— and perhaps contradictory—content, to give everyone involved a broader perspective. Choice of leadership is extremely important, so anything you can do to improve the quality of the decision is probably worth the effort.

In my experience, most people have a somewhat distorted and muddled view of Scrum leadership roles. Unless that distortion is removed, the selection conversations are unlikely to be as meaningful or produce the best result.

16.6   REFERENCE INFORMATION

A variety of chapter-specific reference information is available on the companion website at http://forgingchange.com/fc_sms. This URL has been encoded in the QR code below for your convenience.

Forging Change

Agile Deployment Models

Agile Design Elements

Mgmt. Behaviors in Scrum

Estimating Business Value

Progressive Refinement at Scale

Sprint Alignment Wall

User Story Ruler

Example Scrum Task Boards

Definition of Done Examples

Greatness can't be imposed; it has to come from within. But it does live within all of us.

Jeff Sutherland, *Scrum: The Art of Doing Twice the Work in Half the Time*

Triage Guidelines

Ratcheting Definition of Done

Forecasting Releases

Discerning Genuine Unit Testing

Terminology Definitions

Training Concerns

Scrum Master Selection

Scrum Diagrams

Official Scrum Guide

Virtual Kanban

J. Carpenter

17   Scrum Diagrams

As a wall scrawling analytical type, I find a narrated graphical explanation of Scrum is often very helpful to anyone initially navigating the official Scrum Guide. You might even try to draw these diagrams yourself as you read along — they are very scrawl friendly.

In the event of any discrepancy between the diagrams and the Scrum Guide, the Scrum Guide should take precedence.

17.1   REFERENCE INFORMATION

A variety of chapter-specific reference information is available on the companion website at http://forgingchange.com/fc_sd. This URL has been encoded in the QR code below for your convenience.

Figure 17-1 Scrum overview

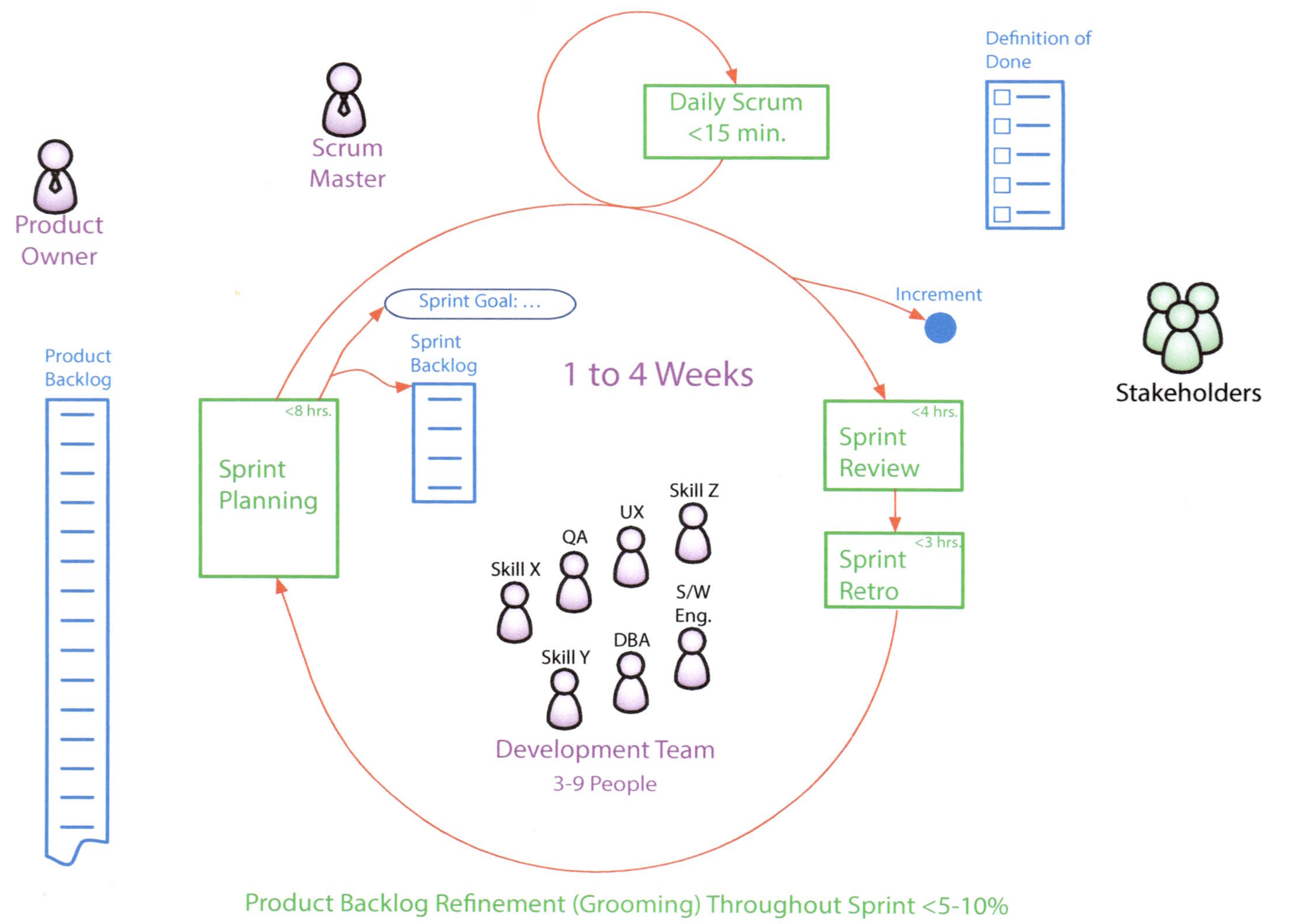

Product Backlog Refinement (Grooming) Throughout Sprint <5-10%

Stakeholders

Definition of Done

Increment

Sprint Review <4 hrs.

Sprint Retro <3 hrs.

Daily Scrum <15 min.

1 to 4 Weeks

Skill Z

UX

S/W Eng.

QA

DBA

Skill X

Skill Y

Development Team 3-9 People

Sprint Goal: ...

Sprint Backlog

Scrum Master

Sprint Planning <8 hrs.

Product Owner

Product Backlog

Forging Change

Agile Deployment Models

Agile Design Elements

Mgmt. Behaviors in Scrum

Estimating Business Value

Progressive Refinement at Scale

Sprint Alignment Wall

User Story Ruler

Example Scrum Task Boards

Definition of Done Examples

Figure 17-2 Product Backlog highlights

Triage
Guidelines

Ratcheting
Definition of
Done

Forecasting
Releases

Discerning
Genuine
Unit Testing

Terminology
Definitions

Training
Concerns

Scrum
Master
Selection

**Scrum
Diagrams**

Official
Scrum
Guide

Virtual
Kanban

J. Carpenter

The **Product Backlog** is a force-ranked list of desired incremental business functionality.

Each Product Backlog Item (PBI) is created as a vertical slice through the entire application infrastructure.

Ideal PBIs meet the INVEST test:
I-Independent
N-Negotiable
V-Valuable
E-Estimatable
S-Small
T-Testable

For a single product there should only be a single Product Backlog regardless of the number of teams working on the product.

Product
Owner

Product
Backlog

Forging
Change

Agile
Deployment
Models

Agile
Design
Elements

Mgmt.
Behaviors
in Scrum

Estimating
Business
Value

Progressive
Refinement
at Scale

Sprint
Alignment
Wall

User Story
Ruler

Example
Scrum
Task Boards

Definition of
Done
Examples

The **Definition of Done** is an explicit quality standard the Development Team is accountable for achieving on each and every Product Backlog Item and associated Increment.

The official Scrum Guide has an excellent longer definition and explanation, which is well worth reading.

It is important to remember the Definition of Done is an **evolving quality standard**. At any point it should represent the best quality the Development Team is currently capable of achieving considering all the constraints, skill sets, and resources available to the team. As the Development Team's skill and capabilities increase, the Definition of Done should be refined. Creating an idealistic, unachievable Definition of Done is counterproductive as it will surely be ignored and become irrelevant.

I recommend ensuring the Definition of Done has a version and date stamp on it.

Carelessly or unilaterally changing the Definition of Done will likely destroy trust within the team and the organization.

It can be useful to think of the Definition of Done as a **rubber stamp with additional acceptance criteria** applicable to every Product Backlog Item.

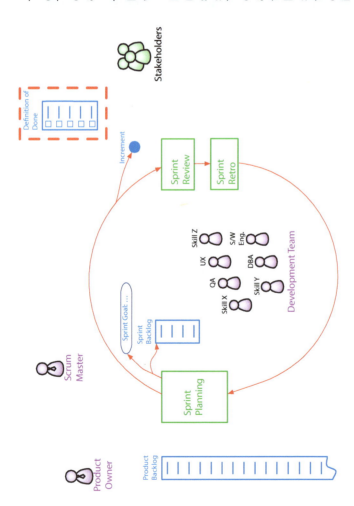

Figure 17-3 Definition of Done highlights

174

Triage
Guidelines

Ratcheting
Definition of
Done

Forecasting
Releases

Discerning
Genuine
Unit Testing

Terminology
Definitions

Training
Concerns

Scrum
Master
Selection

**Scrum
Diagrams**

Official
Scrum
Guide

Virtual
Kanban

J. Carpenter

Figure 17-4 Relationship between Scrum role definitions

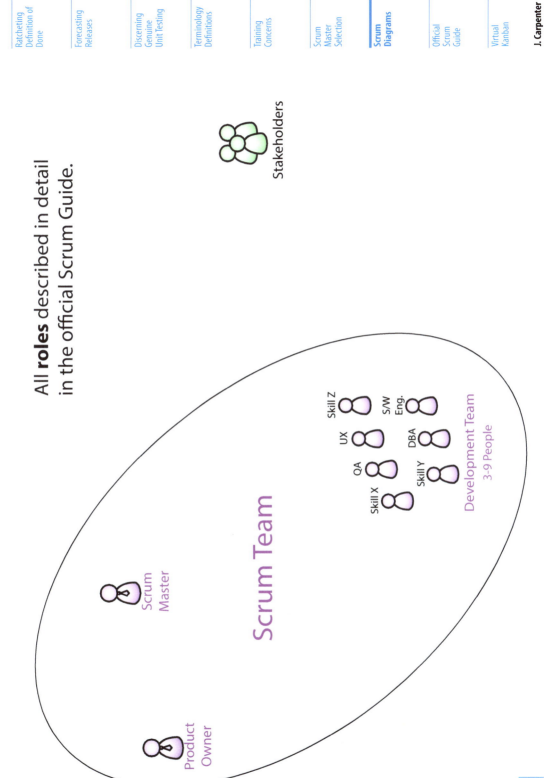

All **roles** described in detail in the official Scrum Guide.

Figure 17-5 Sprint Planning highlights

During **Sprint Planning**, the Development Team and Product Owner work together to figure out a reasonable tactical goal for the Sprint. This tactical goal is captured and articulated into a Sprint Goal.

Based on the Sprint Goal, the Development Team pulls in the Product Backlog Items towards top of the Product Backlog related to the Sprint Goal, and thereby populates the Sprint Backlog (i.e., What).

The Development Team then tasks out the work to the extent possible, primarily as an aid to ensuring effective collaboration, commonality of mind, and visualization of workflow (i.e., How).

The result of Sprint Planning is a **forecast** by the Development team. It is NOT a commitment. The Development Team's commitment is to quality as explicitly expressed in the Definition of Done.

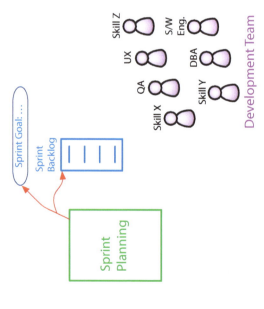

Forging Change

Agile Deployment Models

Agile Design Elements

Mgmt. Behaviors in Scrum

Estimating Business Value

Progressive Refinement at Scale

Sprint Alignment Wall

User Story Ruler

Example Scrum Task Boards

Definition of Done Examples

Triage
Guidelines

Ratcheting
Definition of
Done

Forecasting
Releases

Discerning
Genuine
Unit Testing

Terminology
Definitions

Training
Concerns

Scrum
Master
Selection

**Scrum
Diagrams**

Official
Scrum
Guide

Virtual
Kanban

J. Carpenter

Figure 17-6 Sprint Review highlights

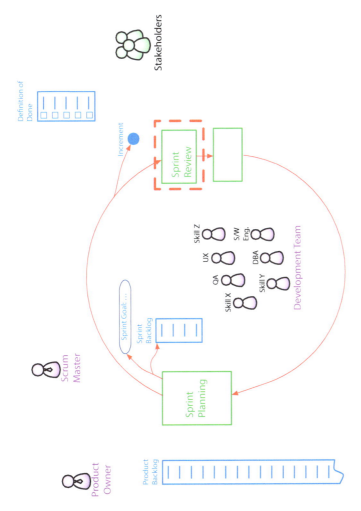

During the **Sprint Review** the Scrum Team demos the DONE Increment to the stakeholders, and conducts a bi-directional conversation with the stakeholders to obtain feedback.

The Scrum Team also talks about the current intentions for the next Sprint and seeks feedback to validate their intended direction.

The Scrum Team also tends to present and obtain feedback regarding their higher-level progress towards larger longer-term product goals.

Generally speaking, the Development Team should NEVER demo Un-Done work. The Increment must be fully shippable. Almost done doesn't count.

The Sprint Review is an **externally focused high-level reflective inspection and adaptation** event.

Forging
Change

Agile
Deployment
Models

Agile
Design
Elements

Mgmt.
Behaviors
in Scrum

Estimating
Business
Value

Progressive
Refinement
at Scale

Sprint
Alignment
Wall

User Story
Ruler

Example
Scrum
Task Boards

Definition of
Done
Examples

Figure 17-7 Sprint Retrospective highlights

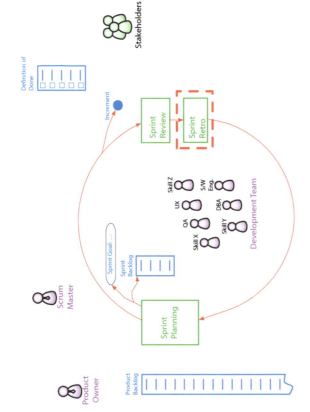

The **Sprint Retrospective** is an opportunity for the Scrum Team to privately reflect upon what has and has not been working, and to devise an actionable plan to improve in the next Sprint.

The most important outcome of the Sprint Retrospective is a concrete improvement goal (i.e., a "try"). Improvement goals should be SMART:

S-specific
M-measurable
A-actionable
R-realistic
T-time bound

When possible, the team should create the improvement goal as a User Story which meets the INVEST test, and agree to place it into the upcoming Sprint's Backlog.

For variety of reasons including personal safety and only involving those accountable for the doing, only the Scrum Team members are typically allowed in a Sprint Retrospective. Whereas the Sprint Review is a formalized high-level external inspection and adaptation event, the Sprint Retrospective is an **internal high-level reflective inspection and adaptation** event.

J. Carpenter

The **Daily Scrum** is a very short tactical coordination meeting by the Development Team, in which the Development Team focuses on what needs done to meet the Sprint Goal and complete the Sprint Backlog.

The following strategy is typically effective, especially for newer teams:

1. Everyone on the Development Team stands in a semicircle around the team's physical task board.

2. Each member very briefly describes:

   a. Progress they achieved since the last Daily Scrum

   b. What they intend to focus on now

   c. Any impediments they are encountering

Standing is a trick to help keep the meeting short. The 3 questions is a trick to help facilitate focus on the goal. Keeping the task board and other information radiators handy tends to keep the team focused on driving to completion while also ensuring the information radiators stay up to date and relevant to the team's needs.

It is very important to keep the meeting short. Identifying the need for any longer discussions and scheduling them is reasonable, holding everyone hostage to a longer discussion is not. The team should learn to self-police their behavior to optimize efficiency.

Typically only the Development Team (and optionally the Scrum Master) are present at the Daily Scrum. Sometimes the very existence of the Product Owner tends to distort the meeting's focus. The Scrum Master does not even need to attend the Daily Scrum for a mature Scrum Team. The Scrum Master is responsible for the ecosystem, not secretarial duties.

Figure 17-8 Daily Scrum highlights

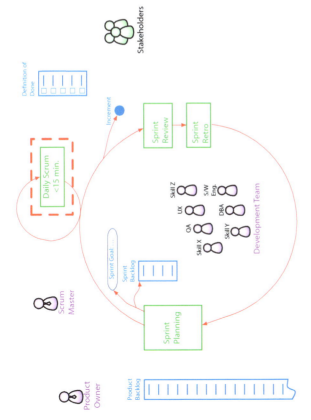

Forging Change

Agile Deployment Models

Agile Design Elements

Mgmt. Behaviors in Scrum

Estimating Business Value

Progressive Refinement at Scale

Sprint Alignment Wall

User Story Ruler

Example Scrum Task Boards

Definition of Done Examples

Throughout the Sprint, the Scrum Team is expected to spend some time **refining the Product Backlog**. The official Scrum Guide does not prescribe how the Product Backlog refinement should be done, just that it should generally consume no more than 10% of the Sprint.

A very common refinement strategy is for Development Team members who know the most about given areas to take a first pass at relevant Product Backlog Items (PBIs) towards the top of the Product Backlog, engaging the Product Owner as needed. The Product Owner and Development Team then have a formal scheduled meeting in which the group further refines the PBIs and assigns relative effort estimates.

WARNING: At all times we must remember to treat estimates purely as estimates, or we will destroy forecast accuracy, morale, and quality. At all times the Development Team's primary obligation is to quality as explicitly articulated in the Definition of Done.

Figure 17-9 Product Backlog refinement highlights

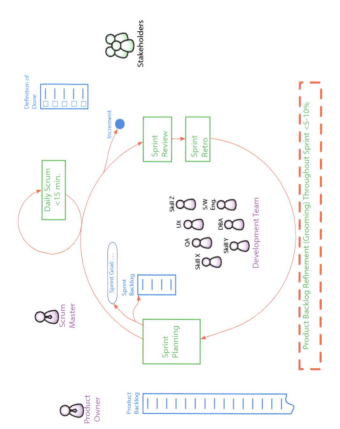

Triage
Guidelines

Ratcheting
Definition of
Done

Forecasting
Releases

Discerning
Genuine
Unit Testing

Terminology
Definitions

Training
Concerns

Scrum
Master
Selection

**Scrum
Diagrams**

Official
Scrum
Guide

Virtual
Kanban

J. Carpenter

Scrum Team Members:

The Scrum Team is composed of the following roles:

Product Owner: Single individual ultimately accountable for **business value**. This includes ensuring the most important functionality is at the top of the Product Backlog.

Scrum Master: Individual ultimately accountable for creating an effective **ecosystem**. This includes ensuring the Development Team has the competency to deliver, and the Product Owner is providing clarity of vision. The Scrum Master can be seen as both a referee and gardener.

Development Team: Ultimately accountable for **Quality** as explicitly expressed in the Definition of Done. The Development Team is solely accountable for delivery.

Stakeholders: Anyone outside of the Scrum Team with a vested interest in the product. This typically includes customers and sponsors of the effort.

Figure 17-10 Scrum role accountabilities

Forging
Change

Agile
Deployment
Models

Agile
Design
Elements

Mgmt.
Behaviors
in Scrum

Estimating
Business
Value

Progressive
Refinement
at Scale

Sprint
Alignment
Wall

User Story
Ruler

Example
Scrum
Task Boards

Definition of
Done
Examples

Scrum is like your mother-in-law, it points out ALL your faults.

Ken Schwaber

Triage
Guidelines

Ratcheting
Definition of
Done

Forecasting
Releases

Discerning
Genuine
Unit Testing

Terminology
Definitions

Training
Concerns

Scrum
Master
Selection

Scrum
Diagrams

**Official
Scrum
Guide**

Virtual
Kanban

J. Carpenter

18   *Official Scrum Guide*

Scrum was invented by Ken Schwaber and Jeff Sutherland. To avoid confusion on what Scrum is and is not, they wrote the official Scrum Guide. Over the years they have made occasional refinements in an effort to ensure a precise understanding of their original intent.

There are many possible process variations based on time-boxed, iterative, and incremental Scrum-like process frameworks, but these variations are not Scrum. Strictly speaking, Scrum is whatever Ken Schwaber and Jeff Sutherland say it is, nothing more and nothing less. This doesn't make non-Scrum empirical process control frameworks good or bad, just not Scrum (and perhaps less well proven).

The Scrum Guide is wonderfully concise. It is intentionally silent in areas Schwaber and Sutherland intend flexibility, and very explicit when they don't. I don't know a better way to begin obtaining a foundation in Scrum than by simply reading the Scrum Guide. As you read it, you are likely to find my diagrams and additional commentary in chapter 17 useful.

The Scrum Guide is offered under the Attribution Share-Alike license of Creative Commons, which allows me to reproduce it here for your convenience. This copy is a direct port of the November 2017 version available as a PDF at: http://scrumguides.org/docs/scrumguide /v2017/2017-Scrum-Guide-US.pdf.

The text has been copied verbatim. I have not moved or deleted a single comma. The only changes have been in typesetting details to fit the book's form factor and typefaces.

The Scrum Guide™

The Definitive Guide to Scrum: The Rules of the Game

November 2017

Developed and sustained by Scrum creators:
Ken Schwaber and Jeff Sutherland

Triage Guidelines

Ratcheting Definition of Done

Forecasting Releases

Discerning Genuine Unit Testing

Terminology Definitions

Training Concerns

Scrum Master Selection

Scrum Diagrams

Official Scrum Guide

Virtual Kanban

J. Carpenter

Table of Contents

Forging Change

Agile Deployment Models

Agile Design Elements

Mgmt. Behaviors in Scrum

Estimating Business Value

Progressive Refinement at Scale

Sprint Alignment Wall

User Story Ruler

Example Scrum Task Boards

Definition of Done Examples

Purpose of the Scrum Guide

Scrum is a framework for developing, delivering, and sustaining complex products. This Guide contains the definition of Scrum. This definition consists of Scrum's roles, events, artifacts, and the rules that bind them together. Ken Schwaber and Jeff Sutherland developed Scrum; the Scrum Guide is written and provided by them. Together, they stand behind the Scrum Guide.

Definition of Scrum

Scrum (n): A framework within which people can address complex adaptive problems, while productively and creatively delivering products of the highest possible value.
Scrum is:

- Lightweight

- Simple to understand

- Difficult to master

Scrum is a process framework that has been used to manage work on complex products since the early 1990s. Scrum is not a process, technique, or definitive method. Rather, it is a framework within which you can employ various processes and techniques. Scrum makes clear the relative efficacy of your product management and work techniques so that you can continuously improve the product, the team, and the working environment.

The Scrum framework consists of Scrum Teams and their associated roles, events, artifacts, and rules. Each component within the framework serves a specific purpose and is essential to Scrum's success and usage.

The rules of Scrum bind together the roles, events, and artifacts, governing the relationships and interaction between them. The rules of Scrum are described throughout the body of this document.

Triage Guidelines

Ratcheting Definition of Done

Forecasting Releases

Discerning Genuine Unit Testing

Terminology Definitions

Training Concerns

Scrum Master Selection

Scrum Diagrams

Official Scrum Guide

Virtual Kanban

J. Carpenter

Specific tactics for using the Scrum framework vary and are described elsewhere.

Uses of Scrum

Scrum was initially developed for managing and developing products. Starting in the early 1990s, Scrum has been used extensively, worldwide, to:

1. Research and identify viable markets, technologies, and product capabilities;

2. Develop products and enhancements;

3. Release products and enhancements, as frequently as many times per day;

4. Develop and sustain Cloud (online, secure, on-demand) and other operational environments for product use; and,

5. Sustain and renew products.

Scrum has been used to develop software, hardware, embedded software, networks of interacting function, autonomous vehicles, schools, government, marketing, managing the operation of organizations and almost everything we use in our daily lives, as individuals and societies.

As technology, market, and environmental complexities and their interactions have rapidly increased, Scrum's utility in dealing with complexity is proven daily.

Scrum proved especially effective in iterative and incremental knowledge transfer. Scrum is now widely used for products, services, and the management of the parent organization.

The essence of Scrum is a small team of people. The individual team is highly flexible and adaptive. These strengths continue operating in single, several, many, and networks of teams that develop, release, operate and sustain the work and work products of thousands of people. They collaborate and interoperate through sophisticated development architectures and target release environments.

When the words "develop" and "development" are used in the Scrum Guide, they refer to complex work, such as those types identified above.

Scrum Theory

Scrum is founded on empirical process control theory, or empiricism. Empiricism asserts that knowledge comes from experience and making decisions based on what is known. Scrum employs an iterative, incremental approach to optimize predictability and control risk. Three pillars uphold every implementation of empirical process control: transparency, inspection, and adaptation.

Transparency

Significant aspects of the process must be visible to those responsible for the outcome. Transparency requires those aspects be defined by a common standard so observers share a common understanding of what is being seen.

For example:

- A common language referring to the process must be shared by all participants; and,

- Those performing the work and those inspecting the resulting increment must share a common definition of "Done".

Triage Guidelines

Ratcheting Definition of Done

Forecasting Releases

Discerning Genuine Unit Testing

Terminology Definitions

Training Concerns

Scrum Master Selection

Scrum Diagrams

Official Scrum Guide

Virtual Kanban

J. Carpenter

Inspection

Scrum users must frequently inspect Scrum artifacts and progress toward a Sprint Goal to detect undesirable variances. Their inspection should not be so frequent that inspection gets in the way of the work. Inspections are most beneficial when diligently performed by skilled inspectors at the point of work.

Adaptation

If an inspector determines that one or more aspects of a process deviate outside acceptable limits, and that the resulting product will be unaccept-able, the process or the material being processed must be adjusted. An adjustment must be made as soon as possible to minimize further devi-ation.

Scrum prescribes four formal events for inspection and adaptation, as described in the Scrum Events section of this document:

- Sprint Planning

- Daily Scrum

- Sprint Review

- Sprint Retrospective

Scrum Values

When the values of commitment, courage, focus, openness and respect are embodied and lived by the Scrum Team, the Scrum pillars of transpar-ency, inspection, and adaptation come to life and build trust for everyone. The Scrum Team members learn and explore those values as they work with the Scrum events, roles and artifacts.

Forging
Change

Agile
Deployment
Models

Agile
Design
Elements

Mgmt.
Behaviors
in Scrum

Estimating
Business
Value

Progressive
Refinement
at Scale

Sprint
Alignment
Wall

User Story
Ruler

Example
Scrum
Task Boards

Definition of
Done
Examples

Successful use of Scrum depends on people becoming more proficient in living these five values. People personally commit to achieving the goals of the Scrum Team. The Scrum Team members have courage to do the right thing and work on tough problems. Everyone focuses on the work of the Sprint and the goals of the Scrum Team. The Scrum Team and its stakeholders agree to be open about all the work and the challenges with performing the work. Scrum Team members respect each other to be capable, independent people.

The Scrum Team

The Scrum Team consists of a Product Owner, the Development Team, and a Scrum Master. Scrum Teams are self-organizing and cross-functional. Self-organizing teams choose how best to accomplish their work, rather than being directed by others outside the team. Cross-functional teams have all competencies needed to accomplish the work without depending on others not part of the team. The team model in Scrum is designed to optimize flexibility, creativity, and productivity. The Scrum Team has proven itself to be increasingly effective for all the earlier stated uses, and any complex work.

Scrum Teams deliver products iteratively and incrementally, maximizing opportunities for feedback. Incremental deliveries of "Done" product ensure a potentially useful version of working product is always available.

The Product Owner

The Product Owner is responsible for maximizing the value of the product resulting from work of the Development Team. How this is done may vary widely across organizations, Scrum Teams, and individuals.

Triage Guidelines

Ratcheting Definition of Done

Forecasting Releases

Discerning Genuine Unit Testing

Terminology Definitions

Training Concerns

Scrum Master Selection

Scrum Diagrams

Official Scrum Guide

Virtual Kanban

J. Carpenter

The Product Owner is the sole person responsible for managing the Product Backlog. Product Backlog management includes:

- Clearly expressing Product Backlog items;

- Ordering the items in the Product Backlog to best achieve goals and missions;

- Optimizing the value of the work the Development Team performs;

- Ensuring that the Product Backlog is visible, transparent, and clear to all, and shows what the Scrum Team will work on next; and,

- Ensuring the Development Team understands items in the Product Backlog to the level needed.

The Product Owner may do the above work, or have the Development Team do it. However, the Product Owner remains accountable.

The Product Owner is one person, not a committee. The Product Owner may represent the desires of a committee in the Product Backlog, but those wanting to change a Product Backlog item's priority must address the Product Owner.

For the Product Owner to succeed, the entire organization must respect his or her decisions. The Product Owner's decisions are visible in the content and ordering of the Product Backlog. No one can force the Development Team to work from a different set of requirements.

The Development Team

The Development Team consists of professionals who do the work of delivering a potentially releasable Increment of "Done" product at the end of each Sprint. A "Done" increment is required at the Sprint Review. Only members of the Development Team create the Increment.

Development Teams are structured and empowered by the organization to organize and manage their own work. The resulting synergy optimizes the Development Team's overall efficiency and effectiveness.

Development Teams have the following characteristics:

- They are self-organizing. No one (not even the Scrum Master) tells the Development Team how to turn Product Backlog into Increments of potentially releasable functionality;

- Development Teams are cross-functional, with all the skills as a team necessary to create a product Increment;

- Scrum recognizes no titles for Development Team members, regardless of the work being performed by the person;

- Scrum recognizes no sub-teams in the Development Team, regardless of domains that need to be addressed like testing, architecture, operations, or business analysis; and,

- Individual Development Team members may have specialized skills and areas of focus, but accountability belongs to the Development Team as a whole.

Development Team Size

Optimal Development Team size is small enough to remain nimble and large enough to complete significant work within a Sprint. Fewer than three Development Team members decrease interaction and results in smaller productivity gains. Smaller Development Teams may encounter skill constraints during the Sprint, causing the Development Team to be unable to deliver a potentially releasable Increment. Having more than nine members requires too much coordination. Large Development Teams generate too much complexity for an empirical process to be useful. The Product Owner and Scrum Master roles are not included in this count unless they are also executing the work of the Sprint Backlog.

Triage
Guidelines

Ratcheting
Definition of
Done

Forecasting
Releases

Discerning
Genuine
Unit Testing

Terminology
Definitions

Training
Concerns

Scrum
Master
Selection

Scrum
Diagrams

**Official
Scrum
Guide**

Virtual
Kanban

J. Carpenter

The Scrum Master

The Scrum Master is responsible for promoting and supporting Scrum as defined in the Scrum Guide. Scrum Masters do this by helping everyone understand Scrum theory, practices, rules, and values.

The Scrum Master is a servant-leader for the Scrum Team. The Scrum Master helps those outside the Scrum Team understand which of their interactions with the Scrum Team are helpful and which aren't. The Scrum Master helps everyone change these interactions to maximize the value created by the Scrum Team.

Scrum Master Service to the Product Owner

The Scrum Master serves the Product Owner in several ways, including:

- Ensuring that goals, scope, and product domain are understood by everyone on the Scrum Team as well as possible;

- Finding techniques for effective Product Backlog management;

- Helping the Scrum Team understand the need for clear and concise Product Backlog items;

- Understanding product planning in an empirical environment;

- Ensuring the Product Owner knows how to arrange the Product Backlog to maximize value;

- Understanding and practicing agility; and,

- Facilitating Scrum events as requested or needed.

Forging
Change

Agile
Deployment
Models

Agile
Design
Elements

Mgmt.
Behaviors
in Scrum

Estimating
Business
Value

Progressive
Refinement
at Scale

Sprint
Alignment
Wall

User Story
Ruler

Example
Scrum
Task Boards

Definition of
Done
Examples

Scrum Master Service to the Development Team

The Scrum Master serves the Development Team in several ways, including:

- Coaching the Development Team in self-organization and cross-functionality;

- Helping the Development Team to create high-value products;

- Removing impediments to the Development Team's progress;

- Facilitating Scrum events as requested or needed; and,

- Coaching the Development Team in organizational environments in which Scrum is not yet fully adopted and understood.

Scrum Master Service to the Organization

The Scrum Master serves the organization in several ways, including:

- Leading and coaching the organization in its Scrum adoption;

- Planning Scrum implementations within the organization;

- Helping employees and stakeholders understand and enact Scrum and empirical product development;

- Causing change that increases the productivity of the Scrum Team; and,

- Working with other Scrum Masters to increase the effectiveness of the application of Scrum in the organization.

Triage Guidelines

Ratcheting Definition of Done

Forecasting Releases

Discerning Genuine Unit Testing

Terminology Definitions

Training Concerns

Scrum Master Selection

Scrum Diagrams

Official Scrum Guide

Virtual Kanban

J. Carpenter

Scrum Events

Prescribed events are used in Scrum to create regularity and to minimize the need for meetings not defined in Scrum. All events are time-boxed events, such that every event has a maximum duration. Once a Sprint begins, its duration is fixed and cannot be shortened or lengthened. The remaining events may end whenever the purpose of the event is achieved, ensuring an appropriate amount of time is spent without allowing waste in the process.

Other than the Sprint itself, which is a container for all other events, each event in Scrum is a formal opportunity to inspect and adapt something. These events are specifically designed to enable critical transparency and inspection. Failure to include any of these events results in reduced transparency and is a lost opportunity to inspect and adapt.

The Sprint

The heart of Scrum is a Sprint, a time-box of one month or less during which a "Done", useable, and potentially releasable product Increment is created. Sprints have consistent durations throughout a development effort. A new Sprint starts immediately after the conclusion of the previous Sprint.

Sprints contain and consist of the Sprint Planning, Daily Scrums, the development work, the Sprint Review, and the Sprint Retrospective.

During the Sprint:

- No changes are made that would endanger the Sprint Goal;

- Quality goals do not decrease; and,

- Scope may be clarified and re-negotiated between the Product Owner and Development Team as more is learned.

Forging Change

Agile Deployment Models

Agile Design Elements

Mgmt. Behaviors in Scrum

Estimating Business Value

Progressive Refinement at Scale

Sprint Alignment Wall

User Story Ruler

Example Scrum Task Boards

Definition of Done Examples

Each Sprint may be considered a project with no more than a one-month horizon. Like projects, Sprints are used to accomplish something. Each Sprint has a goal of what is to be built, a design and flexible plan that will guide building it, the work, and the resultant product increment.

Sprints are limited to one calendar month. When a Sprint's horizon is too long the definition of what is being built may change, complexity may rise, and risk may increase. Sprints enable predictability by ensuring inspection and adaptation of progress toward a Sprint Goal at least every calendar month. Sprints also limit risk to one calendar month of cost.

Cancelling a Sprint

A Sprint can be cancelled before the Sprint time-box is over. Only the Product Owner has the authority to cancel the Sprint, although he or she may do so under influence from the stakeholders, the Development Team, or the Scrum Master.

A Sprint would be cancelled if the Sprint Goal becomes obsolete. This might occur if the company changes direction or if market or technology conditions change. In general, a Sprint should be cancelled if it no longer makes sense given the circumstances. But, due to the short duration of Sprints, cancellation rarely makes sense.

When a Sprint is cancelled, any completed and "Done" Product Backlog items are reviewed. If part of the work is potentially releasable, the Product Owner typically accepts it. All incomplete Product Backlog Items are re-estimated and put back on the Product Backlog. The work done on them depreciates quickly and must be frequently re-estimated.

Sprint cancellations consume resources, since everyone regroups in another Sprint Planning to start another Sprint. Sprint cancellations are often traumatic to the Scrum Team, and are very uncommon.

Triage
Guidelines

Ratcheting
Definition of
Done

Forecasting
Releases

Discerning
Genuine
Unit Testing

Terminology
Definitions

Training
Concerns

Scrum
Master
Selection

Scrum
Diagrams

**Official
Scrum
Guide**

Virtual
Kanban

J. Carpenter

Sprint Planning

The work to be performed in the Sprint is planned at the Sprint Planning. This plan is created by the collaborative work of the entire Scrum Team.

Sprint Planning is time-boxed to a maximum of eight hours for a one-month Sprint. For shorter Sprints, the event is usually shorter. The Scrum Master ensures that the event takes place and that attendants understand its purpose. The Scrum Master teaches the Scrum Team to keep it within the time-box.

Sprint Planning answers the following:

- What can be delivered in the Increment resulting from the upcoming Sprint?

- How will the work needed to deliver the Increment be achieved?

Topic One: What can be done this Sprint?

The Development Team works to forecast the functionality that will be developed during the Sprint. The Product Owner discusses the objective that the Sprint should achieve and the Product Backlog items that, if completed in the Sprint, would achieve the Sprint Goal. The entire Scrum Team collaborates on understanding the work of the Sprint.

The input to this meeting is the Product Backlog, the latest product Increment, projected capacity of the Development Team during the Sprint, and past performance of the Development Team. The number of items selected from the Product Backlog for the Sprint is solely up to the Development Team. Only the Development Team can assess what it can accomplish over the upcoming Sprint.

During Sprint Planning the Scrum Team also crafts a Sprint Goal. The Sprint Goal is an objective that will be met within the Sprint through the implementation of the Product Backlog, and it provides guidance to the Development Team on why it is building the Increment.

Forging Change

Agile Deployment Models

Agile Design Elements

Mgmt. Behaviors in Scrum

Estimating Business Value

Progressive Refinement at Scale

Sprint Alignment Wall

User Story Ruler

Example Scrum Task Boards

Definition of Done Examples

Topic Two: how will the chosen work get done?

Having set the Sprint Goal and selected the Product Backlog items for the Sprint, the Development Team decides how it will build this functionality into a "Done" product Increment during the Sprint. The Product Backlog items selected for this Sprint plus the plan for delivering them is called the Sprint Backlog.

The Development Team usually starts by designing the system and the work needed to convert the Product Backlog into a working product Increment. Work may be of varying size, or estimated effort. However, enough work is planned during Sprint Planning for the Development Team to forecast what it believes it can do in the upcoming Sprint. Work planned for the first days of the Sprint by the Development Team is decomposed by the end of this meeting, often to units of one day or less. The Development Team self-organizes to undertake the work in the Sprint Backlog, both during Sprint Planning and as needed throughout the Sprint.

The Product Owner can help to clarify the selected Product Backlog items and make trade-offs. If the Development Team determines it has too much or too little work, it may renegotiate the selected Product Backlog items with the Product Owner. The Development Team may also invite other people to attend to provide technical or domain advice.

By the end of the Sprint Planning, the Development Team should be able to explain to the Product Owner and Scrum Master how it intends to work as a self-organizing team to accomplish the Sprint Goal and create the anticipated Increment.

Sprint Goal

The Sprint Goal is an objective set for the Sprint that can be met through the implementation of Product Backlog. It provides guidance to the Development Team on why it is building the Increment. It is created during the Sprint Planning meeting. The Sprint Goal gives the Development

Triage
Guidelines

Ratcheting
Definition of
Done

Forecasting
Releases

Discerning
Genuine
Unit Testing

Terminology
Definitions

Training
Concerns

Scrum
Master
Selection

Scrum
Diagrams

**Official
Scrum
Guide**

Virtual
Kanban

J. Carpenter

Team some flexibility regarding the functionality implemented within the Sprint. The selected Product Backlog items deliver one coherent function, which can be the Sprint Goal. The Sprint Goal can be any other coherence that causes the Development Team to work together rather than on separate initiatives.

As the Development Team works, it keeps the Sprint Goal in mind. In order to satisfy the Sprint Goal, it implements functionality and technology. If the work turns out to be different than the Development Team expected, they collaborate with the Product Owner to negotiate the scope of Sprint Backlog within the Sprint.

Daily Scrum

The Daily Scrum is a 15-minute time-boxed event for the Development Team. The Daily Scrum is held every day of the Sprint. At it, the Development Team plans work for the next 24 hours. This optimizes team collaboration and performance by inspecting the work since the last Daily Scrum and forecasting upcoming Sprint work. The Daily Scrum is held at the same time and place each day to reduce complexity.

The Development Team uses the Daily Scrum to inspect progress toward the Sprint Goal and to inspect how progress is trending toward completing the work in the Sprint Backlog. The Daily Scrum optimizes the probability that the Development Team will meet the Sprint Goal. Every day, the Development Team should understand how it intends to work together as a self-organizing team to accomplish the Sprint Goal and create the anticipated Increment by the end of the Sprint.

The structure of the meeting is set by the Development Team and can be conducted in different ways if it focuses on progress toward the Sprint Goal. Some Development Teams will use questions, some will be more discussion based.

Forging
Change

Agile
Deployment
Models

Agile
Design
Elements

Mgmt.
Behaviors
in Scrum

Estimating
Business
Value

Progressive
Refinement
at Scale

Sprint
Alignment
Wall

User Story
Ruler

Example
Scrum
Task Boards

Definition of
Done
Examples

Here is an example of what might be used:

- What did I do yesterday that helped the Development Team meet the Sprint Goal?

- What will I do today to help the Development Team meet the Sprint Goal?

- Do I see any impediment that prevents me or the Development Team from meeting the Sprint Goal?

The Development Team or team members often meet immediately after the Daily Scrum for detailed discussions, or to adapt, or replan, the rest of the Sprint's work.

The Scrum Master ensures that the Development Team has the meeting, but the Development Team is responsible for conducting the Daily Scrum. The Scrum Master teaches the Development Team to keep the Daily Scrum within the 15-minute time-box.

The Daily Scrum is an internal meeting for the Development Team. If others are present, the Scrum Master ensures that they do not disrupt the meeting.

Daily Scrums improve communications, eliminate other meetings, identify impediments to development for removal, highlight and promote quick decision-making, and improve the Development Team's level of knowledge. This is a key inspect and adapt meeting.

Sprint Review

A Sprint Review is held at the end of the Sprint to inspect the Increment and adapt the Product Backlog if needed. During the Sprint Review, the Scrum Team and stakeholders collaborate about what was done in the Sprint. Based on that and any changes to the Product Backlog during the Sprint, attendees collaborate on the next things that could be done to optimize value. This is an informal meeting, not a status meeting, and the presentation of the Increment is intended to elicit feedback and foster collaboration.

This is at most a four-hour meeting for one-month Sprints. For shorter Sprints, the event is usually shorter. The Scrum Master ensures that the event takes place and that attendees understand its purpose. The Scrum Master teaches everyone involved to keep it within the time-box.

The Sprint Review includes the following elements:

- Attendees include the Scrum Team and key stakeholders invited by the Product Owner;

- The Product Owner explains what Product Backlog items have been "Done" and what has not been "Done";

- The Development Team discusses what went well during the Sprint, what problems it ran into, and how those problems were solved;

- The Development Team demonstrates the work that it has "Done" and answers questions about the Increment;

- The Product Owner discusses the Product Backlog as it stands. He or she projects likely target and delivery dates based on progress to date (if needed);

- The entire group collaborates on what to do next, so that the Sprint Review provides valuable input to subsequent Sprint Planning;

- Review of how the marketplace or potential use of the product might have changed what is the most valuable thing to do next; and,

- Review of the timeline, budget, potential capabilities, and marketplace for the next anticipated releases of functionality or capability of the product.

Forging
Change

Agile
Deployment
Models

Agile
Design
Elements

Mgmt.
Behaviors
in Scrum

Estimating
Business
Value

Progressive
Refinement
at Scale

Sprint
Alignment
Wall

User Story
Ruler

Example
Scrum
Task Boards

Definition of
Done
Examples

The result of the Sprint Review is a revised Product Backlog that defines the probable Product Backlog items for the next Sprint. The Product Backlog may also be adjusted overall to meet new opportunities.

Sprint Retrospective

The Sprint Retrospective is an opportunity for the Scrum Team to inspect itself and create a plan for improvements to be enacted during the next Sprint.

The Sprint Retrospective occurs after the Sprint Review and prior to the next Sprint Planning. This is at most a three-hour meeting for one-month Sprints. For shorter Sprints, the event is usually shorter. The Scrum Master ensures that the event takes place and that attendants understand its purpose.

The Scrum Master ensures that the meeting is positive and productive. The Scrum Master teaches all to keep it within the time-box. The Scrum Master participates as a peer team member in the meeting from the accountability over the Scrum process.

The purpose of the Sprint Retrospective is to:

- Inspect how the last Sprint went with regards to people, relationships, process, and tools;

- Identify and order the major items that went well and potential improvements; and,

- Create a plan for implementing improvements to the way the Scrum Team does its work.

The Scrum Master encourages the Scrum Team to improve, within the Scrum process framework, its development process and practices to make it more effective and enjoyable for the next Sprint. During each Sprint Retrospective, the Scrum Team plans ways to increase product quality by improving work processes or adapting the definition of "Done", if appropriate and not in conflict with product or organizational standards.

By the end of the Sprint Retrospective, the Scrum Team should have identified improvements that it will implement in the next Sprint. Implementing these improvements in the next Sprint is the adaptation to the inspection of the Scrum Team itself. Although improvements may be implemented at any time, the Sprint Retrospective provides a formal opportunity to focus on inspection and adaptation.

Scrum Artifacts

Scrum's artifacts represent work or value to provide transparency and opportunities for inspection and adaptation. Artifacts defined by Scrum are specifically designed to maximize transparency of key information so that everybody has the same understanding of the artifact.

Product Backlog

The Product Backlog is an ordered list of everything that is known to be needed in the product. It is the single source of requirements for any changes to be made to the product. The Product Owner is responsible for the Product Backlog, including its content, availability, and ordering.

A Product Backlog is never complete. The earliest development of it lays out the initially known and best-understood requirements. The Product Backlog evolves as the product and the environment in which it will be used evolves. The Product Backlog is dynamic; it constantly changes to identify what the product needs to be appropriate, competitive, and useful. If a product exists, its Product Backlog also exists.

The Product Backlog lists all features, functions, requirements, enhancements, and fixes that constitute the changes to be made to the product in future releases. Product Backlog items have the attributes of a description, order, estimate, and value. Product Backlog items often include test descriptions that will prove its completeness when "Done".

Forging Change

Agile Deployment Models

Agile Design Elements

Mgmt. Behaviors in Scrum

Estimating Business Value

Progressive Refinement at Scale

Sprint Alignment Wall

User Story Ruler

Example Scrum Task Boards

Definition of Done Examples

As a product is used and gains value, and the marketplace provides feedback, the Product Backlog becomes a larger and more exhaustive list. Requirements never stop changing, so a Product Backlog is a living artifact. Changes in business requirements, market conditions, or technology may cause changes in the Product Backlog.

Multiple Scrum Teams often work together on the same product. One Product Backlog is used to describe the upcoming work on the product. A Product Backlog attribute that groups items may then be employed.

Product Backlog refinement is the act of adding detail, estimates, and order to items in the Product Backlog. This is an ongoing process in which the Product Owner and the Development Team collaborate on the details of Product Backlog items. During Product Backlog refinement, items are reviewed and revised. The Scrum Team decides how and when refinement is done. Refinement usually consumes no more than 10% of the capacity of the Development Team. However, Product Backlog items can be updated at any time by the Product Owner or at the Product Owner's discretion.

Higher ordered Product Backlog items are usually clearer and more detailed than lower ordered ones. More precise estimates are made based on the greater clarity and increased detail; the lower the order, the less detail. Product Backlog items that will occupy the Development Team for the upcoming Sprint are refined so that any one item can reasonably be "Done" within the Sprint time-box. Product Backlog items that can be "Done" by the Development Team within one Sprint are deemed "Ready" for selection in a Sprint Planning. Product Backlog items usually acquire this degree of transparency through the above described refining activities.

The Development Team is responsible for all estimates. The Product Owner may influence the Development Team by helping it understand and select trade-offs, but the people who will perform the work make the final estimate.

Triage Guidelines

Ratcheting Definition of Done

Forecasting Releases

Discerning Genuine Unit Testing

Terminology Definitions

Training Concerns

Scrum Master Selection

Scrum Diagrams

Official Scrum Guide

Virtual Kanban

J. Carpenter

Monitoring Progress Toward a Goal

At any point in time, the total work remaining to reach a goal can be summed. The Product Owner tracks this total work remaining at least every Sprint Review. The Product Owner compares this amount with work remaining at previous Sprint Reviews to assess progress toward completing projected work by the desired time for the goal. This information is made transparent to all stakeholders.

Various projective practices upon trending have been used to forecast progress, like burn-downs, burn-ups, or cumulative flows. These have proven useful. However, these do not replace the importance of empiricism. In complex environments, what will happen is unknown. Only what has already happened may be used for forward-looking decision-making.

Sprint Backlog

The Sprint Backlog is the set of Product Backlog items selected for the Sprint, plus a plan for delivering the product Increment and realizing the Sprint Goal. The Sprint Backlog is a forecast by the Development Team about what functionality will be in the next Increment and the work needed to deliver that functionality into a "Done" Increment.

The Sprint Backlog makes visible all the work that the Development Team identifies as necessary to meet the Sprint Goal. To ensure continuous improvement, it includes at least one high priority process improvement identified in the previous Retrospective meeting.

The Sprint Backlog is a plan with enough detail that changes in progress can be understood in the Daily Scrum. The Development Team modifies the Sprint Backlog throughout the Sprint, and the Sprint Backlog emerges during the Sprint. This emergence occurs as the Development Team works through the plan and learns more about the work needed to achieve the Sprint Goal.

Forging Change

Agile Deployment Models

Agile Design Elements

Mgmt. Behaviors in Scrum

Estimating Business Value

Progressive Refinement at Scale

Sprint Alignment Wall

User Story Ruler

Example Scrum Task Boards

Definition of Done Examples

As new work is required, the Development Team adds it to the Sprint Backlog. As work is performed or completed, the estimated remaining work is updated. When elements of the plan are deemed unnecessary, they are removed. Only the Development Team can change its Sprint Backlog during a Sprint. The Sprint Backlog is a highly visible, real-time picture of the work that the Development Team plans to accomplish during the Sprint, and it belongs solely to the Development Team.

Monitoring Sprint Progress

At any point in time in a Sprint, the total work remaining in the Sprint Backlog can be summed. The Development Team tracks this total work remaining at least for every Daily Scrum to project the likelihood of achieving the Sprint Goal. By tracking the remaining work throughout the Sprint, the Development Team can manage its progress.

Increment

The Increment is the sum of all the Product Backlog items completed during a Sprint and the value of the increments of all previous Sprints. At the end of a Sprint, the new Increment must be "Done," which means it must be in useable condition and meet the Scrum Team's definition of "Done". An increment is a body of inspectable, done work that supports empiricism at the end of the Sprint. The increment is a step toward a vision or goal. The increment must be in useable condition regardless of whether the Product Owner decides to release it.

Triage
Guidelines

Ratcheting
Definition of
Done

Forecasting
Releases

Discerning
Genuine
Unit Testing

Terminology
Definitions

Training
Concerns

Scrum
Master
Selection

Scrum
Diagrams

Official
Scrum
Guide

Virtual
Kanban

J. Carpenter

Artifact Transparency

Scrum relies on transparency. Decisions to optimize value and control risk are made based on the perceived state of the artifacts. To the extent that transparency is complete, these decisions have a sound basis. To the extent that the artifacts are incompletely transparent, these decisions can be flawed, value may diminish and risk may increase.

The Scrum Master must work with the Product Owner, Development Team, and other involved parties to understand if the artifacts are completely transparent. There are practices for coping with incomplete transparency; the Scrum Master must help everyone apply the most appropriate practices in the absence of complete transparency. A Scrum Master can detect incomplete transparency by inspecting the artifacts, sensing patterns, listening closely to what is being said, and detecting differences between expected and real results.

The Scrum Master's job is to work with the Scrum Team and the organization to increase the transparency of the artifacts. This work usually involves learning, convincing, and change. Transparency doesn't occur overnight, but is a path.

Definition of "Done"

When a Product Backlog item or an Increment is described as "Done", everyone must understand what "Done" means. Although this may vary significantly per Scrum Team, members must have a shared understanding of what it means for work to be complete, to ensure transparency. This is the definition of "Done" for the Scrum Team and is used to assess when work is complete on the product Increment.

The same definition guides the Development Team in knowing how many Product Backlog items it can select during a Sprint Planning. The purpose of each Sprint is to deliver Increments of potentially releasable functionality that adhere to the Scrum Team's current definition of "Done".

Forging Change

Agile Deployment Models

Agile Design Elements

Mgmt. Behaviors in Scrum

Estimating Business Value

Progressive Refinement at Scale

Sprint Alignment Wall

User Story Ruler

Example Scrum Task Boards

Definition of Done Examples

Development Teams deliver an Increment of product functionality every Sprint. This Increment is useable, so a Product Owner may choose to immediately release it. If the definition of "Done" for an increment is part of the conventions, standards or guidelines of the development organization, all Scrum Teams must follow it as a minimum.

If "Done" for an increment is not a convention of the development organization, the Development Team of the Scrum Team must define a definition of "Done" appropriate for the product. If there are multiple Scrum Teams working on the system or product release, the Development Teams on all the Scrum Teams must mutually define the definition of "Done".

Each Increment is additive to all prior Increments and thoroughly tested, ensuring that all Increments work together.

As Scrum Teams mature, it is expected that their definitions of "Done" will expand to include more stringent criteria for higher quality. New definitions, as used, may uncover work to be done in previously "Done" increments. Any one product or system should have a definition of "Done" that is a standard for any work done on it.

End Note

Scrum is free and offered in this Guide. Scrum's roles, events, artifacts, and rules are immutable and although implementing only parts of Scrum is possible, the result is not Scrum. Scrum exists only in its entirety and functions well as a container for other techniques, methodologies, and practices.

Triage Guidelines

Ratcheting Definition of Done

Forecasting Releases

Discerning Genuine Unit Testing

Terminology Definitions

Training Concerns

Scrum Master Selection

Scrum Diagrams

Official Scrum Guide

Virtual Kanban

J. Carpenter

Acknowledgements

People

Of the thousands of people who have contributed to Scrum, we should single out those who were instrumental at the start: Jeff Sutherland worked with Jeff McKenna and John Scumniotales, and Ken Schwaber worked with Mike Smith and Chris Martin, and all of them worked together. Many others contributed in the ensuing years and without their help Scrum would not be refined as it is today.

History

Ken Schwaber and Jeff Sutherland worked on Scrum until 1995, when they co-presented Scrum at the OOPSLA Conference in 1995. This presentation essentially documented the learning that Ken and Jeff gained over the previous few years, and made public the first formal definition of Scrum.

The history of Scrum is described elsewhere. To honor the first places where it was tried and refined, we recognize Individual, Inc., Newspage, Fidelity Investments, and IDX (now GE Health).

The Scrum Guide documents Scrum as developed, evolved, and sustained for 20-plus years by Jeff Sutherland and Ken Schwaber. Other sources provide you with patterns, processes, and insights that complement the Scrum framework. These may increase productivity, value, creativity, and satisfaction with the results.

Forging
Change

Agile
Deployment
Models

Agile
Design
Elements

Mgmt.
Behaviors
in Scrum

Estimating
Business
Value

Progressive
Refinement
at Scale

Sprint
Alignment
Wall

User Story
Ruler

Example
Scrum
Task Boards

Definition of
Done
Examples

18.1   REFERENCE INFORMATION

A variety of chapter-specific reference information is available on the companion website at http://forgingchange.com/fc_osg. This URL has been encoded in the QR code below for your convenience.

J. Carpenter

Forging Change

Agile Deployment Models

Agile Design Elements

Mgmt. Behaviors in Scrum

Estimating Business Value

Progressive Refinement at Scale

Sprint Alignment Wall

User Story Ruler

Example Scrum Task Boards

Definition of Done Examples

I say an hour lost at a bottleneck is an hour out of the entire system. I say an hour saved at a non-bottleneck is worthless. Bottlenecks govern both throughput and inventory.

Eliyahu M. Goldratt

Triage Guidelines

Ratcheting Definition of Done

Forecasting Releases

Discerning Genuine Unit Testing

Terminology Definitions

Training Concerns

Scrum Master Selection

Scrum Diagrams

Official Scrum Guide

Virtual Kanban

J. Carpenter

19   Virtual Kanban

There is currently a somewhat contentious debate within the agile community over Kanban versus Scrum. David Anderson and others suggest organizations are better off using Kanban techniques to slowly evolve from their current state. In contrast, Scrum's approach presumes the organizational structure is too broken and too entrenched for slow evolutionary change to be effective. Scrum's approach is to radically restructure the organization—effectively rebooting the organization. I believe there are advantages and disadvantages to both approaches. In general I feel Scrum's approach is far more effective at driving positive change, although it causes greater initial disruption.

I think it is a mistake to get too caught up in the Kanban versus Scrum debate. A wiser approach is to use Kanban and Scrum in combination, doing whatever is most appropriate to the situation. Hopefully I can provide insights here which help you make more informed decisions.

19.1   UNDERSTAND KANBAN AND SCRUM BASICS FIRST

This chapter attempts to help you map the relationships between the design elements of Scrum, Kanban, and the agile design elements detailed in chapter 2. Understanding these relationships will provide insights which should help you make better process selection decisions in the future.

A basic introduction to Kanban is beyond the scope of this chapter. As with most of the book, I am primarily focused on clearing up misconceptions and providing real-world, actionable guidance. I doubt anything I could write about the basics would be significantly better than what is already out there.

Forging
Change

Agile
Deployment
Models

Agile
Design
Elements

Mgmt.
Behaviors
in Scrum

Estimating
Business
Value

Progressive
Refinement
at Scale

Sprint
Alignment
Wall

User Story
Ruler

Example
Scrum
Task Boards

Definition of
Done
Examples

If you are completely new to Kanban you may be better off skimming most of this chapter and coming back to it later. In that case, Russell Healy's getKanban simulation game (section 19.5, page 222) and the diagrams found in the chapter-specific references found at http://forging-change.com/fc_vk will probably be the most useful resources for you.

19.2   FLOW CONTROL VERSUS HIGH-LEVEL FEEDBACK LOOPS

Separating the flow control mechanisms of Scrum and Kanban from the higher-level feedback loop mechanisms can provide useful insights. To help you do this, I will set the stage by listing the six general Kanban practices and connecting them to their implementations in Virtual Kanban systems. Then I will connect the dots between the flow control mechanisms of Scrum and Kanban. This will make it easier to contrast the higher-level feedback loops of Scrum and Kanban.

19.2.1   Kanban's Six Practices for Evolutionary Design

In "The Principles & General Practices of the Kanban Method" at http://www.djaa.com/principles-general-practices-kanban-method, David Anderson lists the following six practices for evolutionary design:

1.   Visualize

2.   Limit Work In Progress (WIP)

3.   Manage Flow

4.   Make Process Explicit

5.   Implement Feedback Loops

6.   Improve Collaboratively, Evolve Experimentally

An almost identical list is provided in Wikipedia's "Kanban (development)" entry at https://en.wikipedia.org/wiki/Kanban_(development).

Triage Guidelines

Ratcheting Definition of Done

Forecasting Releases

Discerning Genuine Unit Testing

Terminology Definitions

Training Concerns

Scrum Master Selection

Scrum Diagrams

Official Scrum Guide

Virtual Kanban

J. Carpenter

19.2.2   Kanban Boards Implement the First Four Practices

The first four practices for evolutionary design are generally achieved using physical or electronic Kanban boards. A well-designed Kanban board will have the following features:

- **Visualization of the flow of discrete quantums of work through a set of activity and queue states.** The quantums of work are typically represented as User Stories. The arrangement of activity and queue states on the Kanban board acts as a form of value stream map.

- **Explicit WIP limits.** These are typically defined across both an activity state and the downstream queue state. WIP limits are usually defined in terms of maximum User Story counts or maximum effort estimate sums, and sometimes both.

- **Explicit exit criteria for each activity state.** The union of all exit criteria across all activity states is largely equivalent to Scrum's Definition of Done.

- **One or more input queues containing the prioritized work items (e.g., User Stories).** These work items will sometimes be sorted into multiple classes of service (categories). When there are multiple input queues or classes of service there will generally be explicit pull rules that determine the work item to be pulled in next. The combination of input queues, classes of service, and pull rules can be seen as a form of scheduling algorithm.

19.2.3   Flow Control and Kanban Boards

The design elements of a Kanban collectively act as a pull-based flow control mechanism.

As you may remember, the first three design elements detailed in chapter 2 are Estimates as Estimates, Buffer Management, and Queue Prioritization. These can all be seen as flow control elements.

Forging
Change

Agile
Deployment
Models

Agile
Design
Elements

Mgmt.
Behaviors
in Scrum

Estimating
Business
Value

Progressive
Refinement
at Scale

Sprint
Alignment
Wall

User Story
Ruler

Example
Scrum
Task Boards

Definition of
Done
Examples

The additional design elements detailed in chapter 2 are Fast Interpersonal Feedback Loops, Fast Technical Feedback Loops, and External Customer Focus. None of these relate to flow control.

Revisiting the Kanban board design features and mapping them to the agile design elements, we have these comparisons:

- **Estimates as Elements:** Virtual Kanban's reliance on pull versus push semantics

- **Buffer Management:** The WIP Limits of a Virtual Kanban system

- **Queue Prioritization:** Virtual Kanban's scheduling algorithm mechanisms—prioritized input queues, classes of service, and pull rules

Triage Guidelines

Ratcheting Definition of Done

Forecasting Releases

Discerning Genuine Unit Testing

Terminology Definitions

Training Concerns

Scrum Master Selection

Scrum Diagrams

Official Scrum Guide

Virtual Kanban

J. Carpenter

19.2.4   Flow Control and Scrum

From a flow control perspective, Scrum is an extremely simple Virtual Kanban system. A Product Backlog Item (PBI) in Scrum flows through a value stream map of Not-Started » In Progress » Done. Scrum basically implements a Virtual Kanban system with a single activity state.

You will notice the Scrum Development Team is in complete control of the Sprint Backlog. In other words, the Scrum Development Team "pulls" work from the Product Backlog into the Sprint Backlog. The Sprint Backlog is a forecast, not a commitment. The Scrum Development Team's primary obligation is quality, as explicitly articulated in the definition of Done. The words are different, but this is effectively Estimates as Estimates.

Kanban boards frequently implement more complex scheduling mechanisms by using multiple input queues, classes of service, and explicit pull rules; in contrast, Scrum's scheduling algorithm is trivial. Scrum has a single prioritized input queue (the Product Backlog) without multiple classes of service. Scrum's pull rule is basically "pull from the top of the Product Backlog."

You will notice Scrum Development Teams limit the number of PBIs they pull into the Sprint Backlog based on their judgment of how many they can reasonably forecast. Although not described as such, this is a form of WIP limit.

The Definition of Done in Scrum can be seen as the exit criteria of a PBI's "In Progress" activity state. Whereas a Kanban system may have a complex sequence of activity states, each with their own corresponding exit criteria, Scrum has only one activity state and only one set of exit criteria.

Mapping Scrum flow control mechanisms to the flow control design elements of agile gives us this:

- **Estimates as Elements:** Scrum Development Team commitment to quality as explicitly articulated in the definition of Done—the Sprint Forecast is a forecast, not a commitment.

Forging
Change

Agile
Deployment
Models

Agile
Design
Elements

Mgmt.
Behaviors
in Scrum

Estimating
Business
Value

Progressive
Refinement
at Scale

Sprint
Alignment
Wall

User Story
Ruler

Example
Scrum
Task Boards

Definition of
Done
Examples

- **Buffer Management:** The Sprint Backlog in combination with the Sprint Goal.

- **Queue Prioritization:** Simple forced ranking of the Product Backlog.

So as you can see, from a flow control perspective Scrum is an exceptionally simplistic Virtual Kanban system. A mathematician might say Scrum's flow control is a degenerate form of Virtual Kanban. Simple isn't a bad thing. Scrum's simplicity is part of its elegance.

19.2.5   Higher-Level Feedback Loops

Scrum has lots of well-prescribed feedback loops. Every ceremony in Scrum is an inspection and adaptation event—in other words, a form of feedback loop.

In practice, professional Scrum Teams use a large number of technical software craftsmanship practices such as automated unit testing and static code analysis. They also implement continuous integration and continuous delivery as both a behavior and a tool. These same teams are frequently co-located, and frequently practice pair programming. Although these practices are not explicitly dictated by the official Scrum Guide, they are generally recognized as critical to professional software engineering Scrum Teams. Each of these practices can be seen as a form of feedback loop.

So with Scrum we have a huge list of higher-level feedback loops. In Kanban we have only have practices 5 and 6: "Implement Feedback Loops" and "Improve Collaboratively, Evolve Experimentally."

Triage
Guidelines

Ratcheting
Definition of
Done

Forecasting
Releases

Discerning
Genuine
Unit Testing

Terminology
Definitions

Training
Concerns

Scrum
Master
Selection

Scrum
Diagrams

Official
Scrum
Guide

**Virtual
Kanban**

J. Carpenter

19.2.6   Contrasting Scrum and Kanban

In practice, many of the Kanban feedback loops are quite similar to those used by professional Scrum Teams. Even so, I can't help but feel Scrum to be a far more complete process framework than Kanban's rather lais-sez-faire approach.

In my experience Kanban alone is unable to overcome the political momentum of a legacy organization. Furthermore, whoever is managing the process seldom has sufficient political capital to clamp down the WIP limits enough to drive ideal levels of collaboration between organizational silos.

I am not saying Scrum is the only solution that will work. I am simply saying all the agile design elements listed in chapter 2 must somehow be accounted for. Furthermore, there must be sufficient structural guidance to make it possible to hold middle managers accountable for changing their behaviors.

Forging
Change

Agile
Deployment
Models

Agile
Design
Elements

Mgmt.
Behaviors
in Scrum

Estimating
Business
Value

Progressive
Refinement
at Scale

Sprint
Alignment
Wall

User Story
Ruler

Example
Scrum
Task Boards

Definition of
Done
Examples

19.3   COMPARING KANBAN AND VIRTUAL KANBAN

In a manufacturing context, Kanban systems are often implemented using physical constraints. To limit the WIP of a widget manufacturing machine, a process engineer can instruct the widget machine operator to only produce widgets when there is an empty pallet at the workstation. This prevents completed widgets from stacking up too much while awaiting consumption by downstream machines. If each dedicated widget-holding pallet holds five widgets, the process engineer can limit the widget WIP to ten widgets by ensuring there are only two dedicated widget-holding pallets on the factory floor.

Unlike manufacturing, knowledge work such as software engineering does not deal in physical widgets. The index card or sticky note pulled across a software engineering team's Kanban board is simply a proxy for the work itself. As such a software engineering team's Kanban board is a "Virtual Kanban" system rather than a physical Kanban system.

It is important to carefully discern how anything you read uses the term Kanban. When you read literature from the lean manufacturing domain, *Kanban* probably refers only to Kanban-style flow control mechanisms. When reading in the agile domain, you frequently have to carefully judge whether an author is talking purely about Kanban-style flow control or is using David Anderson's broader definition.

To help avoid confusion, I have tried to use the term Virtual Kanban any time I am limiting my comments to Kanban-based flow control mechanisms for knowledge work. This is consistent with David Anderson's use of Kanban and Virtual Kanban, and should hopefully help to avoid confusion.

When you hear or read about Kanban boards, you can be fairly certain someone is talking about flow control alone.

Triage
Guidelines

Ratcheting
Definition of
Done

Forecasting
Releases

Discerning
Genuine
Unit Testing

Terminology
Definitions

Training
Concerns

Scrum
Master
Selection

Scrum
Diagrams

Official
Scrum
Guide

**Virtual
Kanban**

J. Carpenter

19.4   VIRTUAL KANBAN IS VERY USEFUL

Virtual Kanban techniques are fantastic tools for visualization and flow control of work in the complicated systems domain; they work with a well-understood value stream map. Every large-scale agile implementation has some work which fits this description. Good examples include the progression of a release candidate through a continuous delivery build pipeline and multistage Product Backlog refinement.

The work of implementing a PBI seldom fits a consistent non-trivial value stream map. The typical Scrum Team practice of breaking down a PBI into multiple tasks provides far more flexibility than a rigid value stream allows. Yet if Scrum Teams identify aspects of their work that are better described as complicated than complex, there is nothing wrong with using Virtual Kanban techniques to manage the flow and provide greater insight for those aspects.

19.4.1   Drill Press Analogy

If you have seen Norm Abram's *The New Yankee Workshop* PBS television series, you know he has an amazing woodworking shop filled with every wondrous woodworking tool available. One of these tools is a drill press. It is unlikely that the manual for Norm's drill press says anything about painting, sanding, cutting, or any other operation unrelated to drilling holes. Just because the drill press is limited to drilling holes doesn't make it a bad tool. Norm probably has one of the best drill presses his high-end tool manufacturing sponsors made at the time.

Virtual Kanban is a fantastic tool. Cumulative flow diagrams and statistical control charts are very helpful in understanding complex flows. Large, highly visible Kanban boards are a great way to empower every individual with a detailed overview of the entire system and thereby more efficiently support self-organization.

Just because Virtual Kanban doesn't provide higher-level feedback loops doesn't make it a bad tool. Each tool has its own purpose.

Forging
Change

Agile
Deployment
Models

Agile
Design
Elements

Mgmt.
Behaviors
in Scrum

Estimating
Business
Value

Progressive
Refinement
at Scale

Sprint
Alignment
Wall

User Story
Ruler

Example
Scrum
Task Boards

Definition of
Done
Examples

Oops! Fans of Norm Abram will recognize I have committed a travesty: I mentioned a power tool without a single mention of safety. Norm wishes to correct this. "And remember this: there is no more important safety rule than to wear these—safety glasses."

19.5   RUSSELL HEALY'S GETKANBAN SIMULATION GAME

Russell Healy of getKanban Limited in New Zealand has created a fantastic instructional board game called *getKanban*. You can read more about the game at https://getkanban.com/.

This game is easily the most efficient way I know for people to learn Kanban at the next level of depth. I find players learn in three hours what would otherwise take more than a day to learn at the same level of depth; such is the instructional power of a very effective simulation.

19.5.1   *Cost Considerations*

The newest version of the game is priced at $450 per set. Each set can support one to six players, with three or four being the sweet spot. It is best to play competitively against at least one other team, so ideally you should buy at least two sets.

Long ago, Healy generously made print files for version 2 of the game freely available, and continues to do so. Even the free version costs around $70 per set after you pay for printing costs and supplies. The current version (version 5) is slightly better, yet for many the price difference is hard to justify—especially if you need several game sets.

If funding is not an issue, please consider buying the commercial version. Although $450 per set sounds steep, I'm sure the price is easily justified in terms of Healy's development costs.

Triage
Guidelines

Ratcheting
Definition of
Done

Forecasting
Releases

Discerning
Genuine
Unit Testing

Terminology
Definitions

Training
Concerns

Scrum
Master
Selection

Scrum
Diagrams

Official
Scrum
Guide

**Virtual
Kanban**

J. Carpenter

19.5.2 Producing the Free Version of getKanban

The chapter-specific information at http://forgingchange.com/fc_vk contains instructions which will greatly accelerate any effort to produce your own copy of the free version of the game.

19.6   ELABORATE KANBAN BOARD EXAMPLE SEQUENCE

An online search will quickly find a large number of examples demonstrating how cards are pulled across a simple Kanban board. Finding a detailed example with step-by-step card flow for a more sophisticated example is often much harder. Those who already understand card flow across basic Kanban boards will hopefully find my more sophisticated example in the chapter-specific reference information useful.

19.7   REFERENCE INFORMATION

A variety of chapter-specific reference information is available on the companion website at http://forgingchange.com/fc_vk. This URL has been encoded in the QR code below for your convenience.

Notes

1. Davies, Rick. "Cynefin Framework versus Stacey Matrix versus network perspectives." Rick On the Road. August 20, 2010. Accessed December 13, 2017. http://mandenews .blogspot.com/2010/08/test3.html.

2. Pink, Daniel. *Drive: The Surprising Truth about What Motivates Us*. New York: Riverhead Books, 2009.

3. Reinertsen, Donald. *The Principles of Product Development Flow: Second Generation Lean Product Development*. Redondo Beach, Calif: Celeritas, 2009.

4. Heusser, Matthew. "Using Weighted Shortest Job First (WSJF) to prioritize your backlog and improve ROI." *TechBeacon*. Accessed January 15, 2018. https://techbeacon .com/prioritize-your-backlog-weighted-shortest-job-first -wsjf-improved-roi.

5. Patton, Jeff. *Jeff Patton & Associates*. Accessed January 15, 2018. https://jpattonassociates.com/.

6. Pichler, Roman. *RomanPichler*. Accessed January 15, 2018. https://www.romanpichler.com.

www.ingramcontent.com/pod-product-compliance
Lightning Source LLC
Chambersburg PA
CBHW041428050326
40690CB00002B/469